WARGAMES
TERRAIN & BUILDINGS

WWI TRENCH SYSTEMS

WARGAMES
TERRAIN & BUILDINGS

WWI TRENCH SYSTEMS

Douglas Hardy

Pen & Sword
MILITARY

AN IMPRINT OF PEN & SWORD BOOKS LTD.
YORKSHIRE ~ PHILADELPHIA

First published in Great Britain in 2021 by
Pen & Sword Military
An imprint of
Pen & Sword Books Ltd
Yorkshire – Philadelphia

Copyright © Douglas Hardy, 2021

ISBN 978 1 52679 354 6

The right of Douglas Hardy to be identified as Author of this work has been asserted by him in accordance with the Copyright, Designs and Patents Act 1988.

A CIP catalogue record for this book is available from the British Library.

All rights reserved. No part of this book may be reproduced or transmitted in any form or by any means, electronic or mechanical including photocopying, recording or by any information storage and retrieval system, without permission from the Publisher in writing.

Typeset in Adobe Caslon Pro 10/13 by SJmagic DESIGN SERVICES, India.
Printed and bound in India by Replika Press Pvt. Ltd.

Pen & Sword Books Limited incorporates the imprints of Atlas, Archaeology, Aviation, Discovery, Family History, Fiction, History, Maritime, Military, Military Classics, Politics, Select, Transport, True Crime, Air World, Frontline Publishing, Leo Cooper, Remember When, Seaforth Publishing, The Praetorian Press, Wharncliffe Local History, Wharncliffe Transport, Wharncliffe True Crime and White Owl.

For a complete list of Pen & Sword titles please contact

PEN & SWORD BOOKS LIMITED
47 Church Street, Barnsley, South Yorkshire, S70 2AS, England
E-mail: enquiries@pen-and-sword.co.uk
Website: www.pen-and-sword.co.uk

Or
PEN AND SWORD BOOKS
1950 Lawrence Rd, Havertown, PA 19083, USA
E-mail: Uspen-and-sword@casematepublishers.com
Website: www.penandswordbooks.com

CONTENTS

Introduction		7
Chapter 1	Wargaming the Great War	13
Chapter 2	Materials	15
Chapter 3	Commercial or Scratch-Built?	24
Chapter 4	Building and Painting Commercial Scenery	27
Chapter 5	Scratch-Building	74
Chapter 6	Projects	96
Chapter 7	Researching First World War Trenches Today	104

INTRODUCTION

'The trenches' in many ways define how we think about the Great War. They are also the way in which the armies of the time described the war. British soldiers at that time probably wouldn't have called it the Western Front (it was really only a Western Front to the German army as they also had an Eastern Front) but simply France, the Front or, more likely, the trenches. The war spread worldwide, so as well as on the Western Front there were trenches in Gallipoli, on the vast Eastern Front, in the mountains of Italy, the heat of East Africa, the plains of Salonika and the deserts of Mesopotamia and Palestine. On almost every front of the war the trench was to be found.

The aim of this book is to give you a guide to the techniques for making and painting scenery for your First World War wargames, setting out step-by-step how to make your scenery and then get it ready for your tabletop battles in a clear and easy-to-follow way.

I will look at the materials you will need, the advantages and disadvantages of both commercially available scenery and building things yourself, the different scales and styles of game you can use – some of which will of course depend on which wargames rules you are using – and some examples of trenches that you can buy. I will also cover some projects for adding scenery pieces to enhance your wargames table and create a period atmosphere.

Trenches in Past Wars
Trenches have been a feature of siege warfare since medieval times and can be found in many accounts of more recent battles; for example, in the sieges of Badajoz during the Peninsular Wars where trenches were built to protect the artillery. With the increase in industrialized warfare in the second half of the nineteenth century, trenches became much more prevalent and were used in the Crimean War, notably in the siege of Sevastopol in 1854–55 and in the later stages of the American Civil War, especially around Petersburg in 1864–65. The trenches around Petersburg are often said to be the forerunners of those on the Western Front.

In later years in that century primitive trenches featured in the Boer Wars, where small numbers of Boers were able to withstand attacks by much larger British armies and cause considerable casualties. The same was true of the rocky sangars used by the Pathans in Afghanistan. In the early twentieth century trenches also featured in the Russo-Japanese war of 1904–1905 and in the various Balkan wars that broke out in the years just before the First World War.

Of course, trench warfare did not stop with the end of the Great War. There were trenches to some extent in the Spanish Civil War of the 1930s and to a very large extent in some parts of the Second World War, such as the defences on the Normandy coast and on the Eastern Front. So the techniques described in this book are not limited to the First World War; you can use them for any period where trenches could feature as a part of your game.

Trench Warfare in the First World War

However, it is with the First World War that trenches are most closely associated. In the west, trench warfare began in earnest with the Battle of the Aisne in September 1914 when the German army, having failed in its attempt to sweep behind the Allied armies and end the war in the west in forty days, pulled back to high ground and began to dig in. Yet in fact trenches in some form had been around since the very start of the war. Even at Mons there are accounts of the British army being in 'firing trenches' on the front line, although of course these were not the sort of trenches we now think of; they were more like what we would call today foxholes or rifle pits. In what became known as the Race to the Sea, with each army trying to outflank the other and failing, the trench systems gradually spread like cobwebs to the Channel coast at Nieuwpoort. At the same time they were moving in the other direction too, going through the Vosges Mountains to the Swiss border. By November 1914 the 'Western Front' had established itself across Belgium and France.

Each side thought of their trenches in slightly different ways. The Germans, having captured sizeable chunks of France and most of Belgium, not unreasonably wanted to hang on to it for as long as they could and so mostly saw their trenches as defensive. The Allies, especially the French, were equally keen to shove the Germans out and so saw their trenches more as temporary places to protect their troops while they prepared to take the offensive.

Trenches developed over the course of the war to eventually become formidable defensive structures. At first they were nothing like the trenches that we now think of, but were little more than narrow pits a couple of feet across with very limited traverses, if any. It was the Germans who began building the first comprehensive trench systems, based on their fieldwork manuals and overseen by their engineers. By the end of 1914 fairly basic trenches had been built and connected with each other. They were usually fully manned at this stage of the war as, especially on the Allied side, they were regarded as giving protection for whole units. This made the most of their firepower, but also exposed the whole unit to a high degree of risk.

With the increased use of artillery, the Germans took the lead in developing the trench systems and by early 1915 they were building successive lines of trenches, linked to each other by communication trenches. This principle was quickly taken up by the Allied armies too. The Germans in particular also developed a system of multiple lines of defence, with a second and sometimes a third line of trenches a mile or two behind the first.

Each trench system tended to have three (or more) inter-connected lines. The first or front line would be a fire trench with firing bays and traverses to minimize the explosive damage from shells and also to provide a means of defence should the enemy capture a section of the line. Behind the firing trench would be a support line, connected to it by communication trenches. This line might also have firing positions, but was designed so that troops could quickly support their comrades in the front line if it was under attack and would probably be where the company command dugouts were situated. It was also sometimes known as a supervision line, to allow commanders to move quickly between areas of the front line.

Behind this line, and also connected, would come the reserve line where battalion command dugouts and aid stations would be. In the firing and support lines it was quite common for strongpoints – variously known as keeps or redoubts – to be built with overlapping fields of fire. Each trench would have a parapet at the front and a parados at the back to provide protection from shells bursting nearby.

Another well-used feature of a trench was the sap. This was a trench dug forward from the main trench towards the enemy, which could be used as a jumping-off point for trench raids or attacks, as a listening post to detect enemy activity or as a forward rifle or machine-gun position. Saps could also be used as a way of edging a trenchline forward by digging two saps and then joining them up. This was a tactic often used by the British to move gradually forward to the German lines or as a jumping-off point for an offensive. The French also frequently made use of something called Russian saps, which were basically unsupported tunnels which then came up in the middle of no man's land.

It wasn't always practical to dig the trenches into the ground, of course. In areas such as the Ypres salient and northern Flanders where the water table was always high, trenches would frequently be built fully above ground and composed of heaps of earth topped by sandbags, though they would often still be half-full of water. Similarly in rocky areas such as the Vosges Mountains, the mountains of Italy and some parts of Gallipoli the trenches were hewn through the rock.

Then there was the question of mud. Some of the classic images of the Great War are of flooded trenches and soldiers stumbling forward through a morass, or balancing precariously on duckboards as they threaded their way through a shell-shattered wood. There were certainly occasions when it was exactly like this: for example, towards the end of the Somme campaign in late autumn 1916 and in the valleys in front of the Passchendaele Ridge in 1917, but in reality these were very much the exception rather than the rule. Until 1916 there simply wasn't the volume of artillery to really destroy the land in this way, and in any event certainly in the British army many of the shells fired were airburst shrapnel that would not have this effect on the terrain. Millions of shrapnel balls are still in the landscape today and I have managed to pick some up on almost every visit I have made to the Western Front.

British shrapnel balls.

From 1916 to the end of the war artillery barrages increased in intensity and would very much affect the terrain, producing the shell-shattered wasteland that is symbolic of the war, but this was not the case everywhere. One of the reasons that the Cambrai offensive was launched where it was in 1917 was that the ground was 'suitable for tanks'; in other words it wasn't yet churned up by artillery. Tanks had previously been used in the Ypres salient in 1917 but had been a disastrous failure as they simply floundered in the muddy conditions.

For much of the war the area between the lines, or no man's land, was often in fact an area of overgrown grass and weeds. It was like this for almost all the 1915 battles on the Western Front, and on 1 July 1916 the British infantry advanced through knee-high grass on the Somme. It stayed that way for much of the offensive until the ground became really cut up in October and November with heavy rain. There is also a story about a Scottish battalion in 1915 who one morning observed about twenty German soldiers climb out of their trench and begin to mow the long grass in front of it. They were not fired at as the Scots commanders wanted the grass cut anyway since it could hide enemy soldiers!

You will also have seen the many photographs of totally destroyed towns and villages, and read accounts of places being reduced by heavy artillery to mere smears of brick dust in the mud. This was certainly true of farms and villages in some areas from late 1916 onwards, but not everywhere or all of the time. Before then, buildings just behind the front lines could be ruined or sometimes very nearly intact and often served as rest areas for troops just behind the trenches. There are accounts of soldiers watching artillery barrages of enemy trenches from the upper floors of part-ruined buildings, and of course a ruin makes an ideal sniper position, for which they were frequently used. The Germans also made great use of cellars, fortifying them for use as machine-gun positions or building them into redoubts. In fact the Germans were adept at making use of any natural or man-made position to strengthen their defences, especially features such as quarries.

By 1916 the trenches had developed into the classic trench systems that we would recognize now, but they were continuing to evolve. The Germans were working on a strong line of defence that they knew as the Siegfried Stellung and the Allies called the Hindenburg Line, which they had built from just south of Arras to Soissons in the south. In early 1917 they withdrew to this line, laying waste to all the land in between. These trenches incorporated the increasing use of concrete and many bunkers and dugouts. In the north around Ypres bunkers were also increasingly being used, both as firing positions and shelters (known to the Germans as *Stollen*). During the Passchendaele campaign, when the drainage system had been completely destroyed by incessant artillery fire and made worse by heavy rain, there were effectively no trenches at all, merely lines of pillboxes with roughly connected shell-holes forming the defensive lines.

It wasn't only on the British part of the front that bunkers and pillboxes were used, especially but not exclusively by the Germans. On the Chemin des Dames in 1917 the German army had built up an extensive network of defensive systems including a large number of pillboxes. They also made use of concrete *Stollen* in the second and third lines so that counter-attacking forces could shelter from the inevitable bombardment. The Germans had been on the Chemin des Dames for more than two years and had previously regarded it as a quiet or 'rest' area for troops that had been badly mauled in action at Verdun or on the Somme. They had ample time to prepare very strong defences, which the French found impossible to crack in their ill-fated 1917 Nivelle offensive.

Introduction

By 1918 it wasn't just the German army that had developed a defensive strategy. The British knew that they would very likely be facing a major German offensive somewhere in 1918, and in the early part of that year set about preparing their defences to meet it. This wasn't easy, as many of the positions they were in were simply where an offensive had reached the previous year, not necessarily the best place to defend. They set about building a defensive system of three zones. The trenches in the forward zone would only be a lightly-held area, just enough to make the Germans attack it, and then the idea was that the troops would withdraw to the main battle zone, where a network of trenches and redoubts would hold off the German attacks. Behind this would be a reserve zone, though in practice this was often not finished, or sometimes not even started, by the time the Germans attacked in March 1918. In his book *The Kaiser's Battle* on the 1918 *Kaiserschlacht* offensive, Martin Middlebrook gives a good description of the difficulties facing the infantry preparing for the battle, quoting a company commander in the Fifth Army:

'We took over the right sub-section of the front line from the 6th Connaught Rangers. I was very worried about the scanty way in which the front line was held. The trenches were poor and shallow and the wiring in front consisted of a single strand of barbed wire held by screw-iron stakes here and there and, in stretches, this was on the ground forming no defence at all..... I complained bitterly of our weakness, but nothing could be done and we had to put up with it.'

The various armies tended to use different ways to support and reinforce their trenchlines. The British in particular favoured sandbags, and often supported their trench sides with planking and corrugated iron sheets, or 'wrinkly (or sometimes wriggly) tin' as it was known. This often also formed makeshift roofs or shelters to provide some cover from the elements, though not much from incoming fire. The French and the Germans, while also using sandbags and planking (though the French made less use of sandbags on their trench sides), often formed their trench sides from a kind of wattling as they felt this was more flexible and gave better support if the trench side was caved in by fire. The French were generally thought of by other troops as having not very well made trenches: they were much shallower than others and less well revetted. British troops taking over sections of the line from the French often complained about the poor state of the trenches they were moving into and had to put quite a lot of work into making them more robust.

As the war evolved, so did the flooring for trenches too. Both the British and the Germans developed varieties of an inverted A-frame system, with the duckboards at the bottom raised above ground level so that the troops kept their feet comparatively dry. This wasn't just a comfort issue, as the dramatic rise in the numbers of soldiers going down with 'trench foot' due to having permanently wet feet was causing a great deal of concern on both sides. This was naturally worse in some areas – such as Flanders – than others, but was a problem right across the Western Front.

At various points in the trenches there would also be some sort of crossing over the top of the trench. This would often just be a few planks laid across the top of the trench to allow troops who were 'over the top' – perhaps at night on a patrol from the second line – to cross the trench without having to go down into it. Such crossing-points often feature in photographs of trenches at the time. Some trenches also had areas known as 'elephant shelters' which were parts of the trench covered by wrinkly tin. This didn't have much of a protective capability as a shell would easily go through wrinkly tin, but did perhaps provide some shelter for the troops.

In front of the trenches there would generally be a system of barbed wire. In the early days this could be limited to a few strands of wire or quite often wire on 'knife rests', wooden constructions that could quickly be put in place, and these were also used as trench blocks to defend a section of the line if this was needed. The early barbed-wire systems required the troops to hammer posts into the ground to support them, which was unpopular as this activity was largely done at night and the sound could be almost guaranteed to attract enemy fire. By mid-war these were replaced with metal rods with a screw on the end which could quickly, and quietly, be inserted into the ground.

As the war developed the belts of barbed wire became more numerous and thicker. British and French wire tended to be in sections wound between posts in a maze of wire and about 4 or 5ft high. The Germans began using coiled wire, which could often be many feet thick and up to 8 or 10ft high. This was very difficult to cut through, and one of the early functions of tanks was to crush the German wire systems so that the troops could make their way through. At Cambrai in 1917 some tanks had attachments on them to pull the piles of barbed wire out of the way of advancing troops. British tanks were designed to cross trenches, though the Germans began to widen their trenches after they first saw tanks in 1916. This was not the case with the French tanks, however, as both the Schneider and St Chamond tanks, which were based on the Holt Tractor system, found trench-crossing very difficult and they were much more likely to bog down and become stuck.

Wire entanglements were of course frequently destroyed or damaged by shellfire, although in some cases it only tended to make them more entangled and even less able to be penetrated. German wire was also a thicker gauge than Allied wire, with larger and more frequent barbs.

Trenches in theatres of the war other than the Western Front were broadly similar. From photographs of the time the Russians and Austro-Hungarians both seem to have used a wattling system of some sort, though the Russian trenches were famously poorly constructed on the whole. In Gallipoli the Turks appear to have used overhead log protection on their trenches in places; something that was not often seen on other fronts but could make an interesting feature on the wargames table. Trenches in Palestine and Mesopotamia look like they followed the more basic trench systems at the start of the war, but did develop along similar lines to those on the Western Front as the war went on.

Chapter 1

WARGAMING THE GREAT WAR

Of all the periods in wargaming, it is fair to say that the Great War has until recently been somewhat neglected, but with more manufacturers bringing ranges of figures out and more sets of rules gradually being published, the period is starting to gain in popularity. No doubt the recent centenaries have helped with this.

The main problem with gaming the Great War was that, for a long time, no one saw a game in it. There were no flanks to turn, no cavalry dashing about (well, not after October 1914 anyway, although they were in fact used quite often) and no scope for tactical genius, or at least that's what people thought. The rules that were around until recently tended to focus either on gaming at the skirmish level, where one figure on the table equals one man in real life, or more abstract gaming at the divisional level where a couple of bases of about ten figures or so each represents a battalion of around 1,000 men. Recent sets of rules have redressed this imbalance, but in comparison with other more 'popular' periods there are far fewer sets of rules available. In 15mm Flames of War have a set for the later part of the Great War, and for bigger scales the Lardies have their set 'Through the Mud and the Blood'. Warhammer Historical published a set on the Great War, though these are now very hard to obtain, and for higher-level games there are sets such as 'Bloody Picnic', 'Over the Top for Command Decision' and 'Great War Spearhead'.

To succeed in a game you need to use the right tactics, as in the real thing. In the second half of the war at least, get your artillery plan right, use aircraft if you can to spot their artillery and counter-battery it, get the chaps across no man's land while keeping the enemy's heads down and you'll be fine!

Trenches in Wargaming
If you want to wargame the First World War, you will at some stage almost inevitably find yourself needing a trench system, especially if you want to play games in the period after the first month or so of the war.

A trench raid is a popular scenario for platoon-level games, as a raid could be carried out by as few as twenty or so men or on occasions by as many as a whole battalion. Both sides on the Western Front actively raided throughout the war, although the French were not as keen on doing so as the British or Germans.

The first raid of the war took place in daylight in November 1914, when Captain Forrester of the Black Watch led about twenty men on raid on a German machine-gun position in a trench near La Bassée. A second attack took place at night a few days later, when 100 men from the Garhwal Rifles attacked German trenches.

Raiding did not become properly organized until 1916, however: between December 1915 and May 1916 the BEF raided the Germans sixty-three times, forty-seven of which were regarded as 'successful'. By the time of the Somme offensive, between July and November 1916, this had risen to a total of 310 times. The Germans were not to be outdone, however, and hit back with more than sixty-five raids in the same period. Raiding continued in this tit-for-tat way throughout the war.

In his book *Battles with Model Soldiers*, the late Donald Featherstone, often regarded as the father of wargaming, suggested that greater atmosphere could be gained for a night trench raid game if it was played in darkness, with only a torch shone randomly at the ceiling every so often to represent one side or the other sending up a parachute flare! Personally I have never tried this, but am very tempted to give it a go just to see how it plays!

Games set at the battalion level or above will generally represent an attack by one side with the other defending their trenches. It is in these games that factors such as artillery barrages and the ability to cross no man's land with your reserves become critically important. Using some of the rule sets that represent a larger playing scale, it would be possible to play scenarios that cover some of the major offensives of the war.

Scales

First World War wargames figures are available in almost every scale that wargamers use. There are manufacturers producing them in 6mm, 10mm, 15mm, 20mm and 28mm, with a variety of ranges covering the majority of the conflict. It is not easy to say which scale has the most advantages as all have their merits and while many of you will be firmly wedded to one scale, there are many of you out there (I know!) who will have figures in more than one – and sometimes every available – scale.

Ready-built trenches can be purchased in many of these scales, plus of course the increasing use now of 3D printing which will soon make scenery in almost any scale readily available to many wargamers.

Chapter 2

MATERIALS

In this Chapter I am going to look at the various materials you will need to create your trench models. Most will be readily available from your hobby or DIY store, or online.

Styrofoam

Styrofoam is one of the most useful materials for building many types of scenery. It comes in various thicknesses and can easily be cut and sculpted into any manner of shapes. Due to the density of the material it combines the strength needed to make terrain with a light weight to make it easily transportable if necessary.

If you are building trench baseboards you will need it in varying thicknesses depending on the scale of the figures you are using and the size of the boards you want to build. The 5mm and 10mm thicknesses are really useful sheets as you can use them for a number of projects including building ruined houses, adding fire-steps to trenches and making emplacements. For building your trench baseboards you will need either 20mm, 30mm or even 50mm thick if you are using 28mm figures.

I have always bought our Styrofoam online from various suppliers. The sheets, which are also often used in insulation, used to be available in both blue and green but some appear now to have changed to a grey or charcoal black. The colour they come in is irrelevant since you will be painting them in any case!

MDF Board

Useful for basing the sheets of Styrofoam to give them strength when you are building trench boards and for basing your stand-alone terrain pieces. You could also use hardboard for this purpose if you wish.

Coffee-Stirrers

One of the most useful pieces of kit for building and improving trench models, you will find yourself needing hundreds (or perhaps thousands!) of these for any number of different uses. You can buy them online (do not steal them from your local coffee shop; you will need far too many of them for that!) in various quantities and widths.

Matchsticks

Another staple of trench-building and also something you will need in large quantities. Again easily buyable online or in hobby stores. It is best to buy the ones that are simply matchstick-shaped, not real matches as you'll quickly find striking all those matches and cutting their heads off very tiresome!

Materials

Cocktail Sticks

You will find these very handy for supporting areas of Styrofoam while it glues, and also for making posts for barbed-wire systems.

String

String is surprisingly useful for some First World War trench scenery. As the Germans and to some extent the French used wattle on many of their trench sides, string is an easy and quick way to represent this on your trench model. It's best to use a fairly thin string as you don't want it to look too thick.

Glues

You will need a range of different glues and adhesives. As Styrofoam is dissolved by solvents, you can't use any solvent-based adhesives on it (unless you want to get an effect by doing that!). Tubes of solvent-free adhesive are readily available in most DIY stores and you'll find

that it's something that you use quite a bit. It sticks fairly quickly but not instantly, so some things may need to be supported while they dry, but it's a pretty forgiving glue and you can mess about with things for some time before it has stuck for good. It will also stick most things, though if the surface is very absorbent you may need to use a bit more of it.

For some items such as the vacuum-formed (hereafter vac-formed) scenery you can use glues such as UHU. The disadvantage of UHU is of course the stringiness of it, but it will stick things pretty well once it has cured. There is a new form of UHU called All Plastics which I have used to stick resin kits and even soft plastic figures together which has worked quite well. There is also a solvent-free form of UHU.

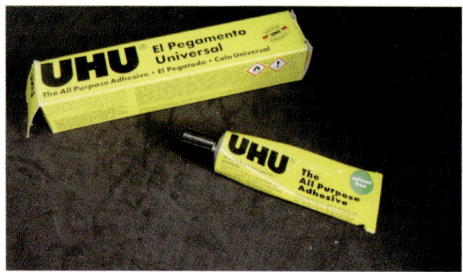

PVA glue is a favourite standby of scenery model-makers everywhere. I tend to buy ours from hobby shops in 5-litre quantities and then decant what I need as it is so useful. It is also solvent-free, but I have learned from experience that it doesn't stick Styrofoam anywhere near as well, or as permanently, as solvent-free adhesive does. It is also slower to cure than solvent-free adhesive.

Superglue in its various forms is very much solvent-based so don't let it anywhere near your Styrofoam! It can be useful, however, for sticking barbed wire onto posts as it will hold quickly so you can get the tension right between posts.

Knives and Chisels

A sharp craft knife is a must for cutting and shaping Styrofoam, and lots of other uses. When you are carving the foam you will need the blade to be extended from the handle so do please be very careful! You are meant to be carving the foam, not your fingers, and it is very easy to slip. You could also use hot wire-cutters as an alternative.

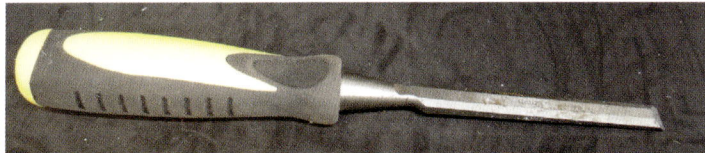

Chisels are also really useful sometimes when carving out shapes in the Styrofoam, especially for building shell-holes.

Scissors

As well as obvious uses such as cutting the string that you are using to represent the wicker-worked sides of your trenches, scissors are also useful for cutting up lengths of coffee-stirrers (this will be very repetitive!) and even trimming model barbed wire.

Modelling Clay

Air-dried modelling clay is the most useful and cheapest way I have found to represent another iconic item of the trench war: sandbags. You will find that you can mass-produce these very easily in an evening or two, and you will always find uses for them. Of course, being air-dried clay you also need to stick them in place quite quickly too. You could use other modelling products such as Milliput, but I have found this to be the most effective method.

Corrugated Iron

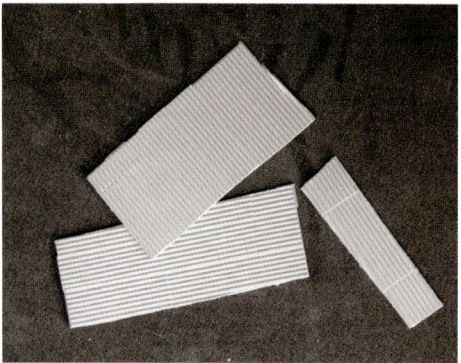

Sheets of plastic corrugated iron, or wrinkly tin as it was often known at the time, are really handy when building your trenches. They are available from model railway suppliers in a variety of sizes and can be cut and shaped as you need them.

Paint

Whatever model you buy or build you are going to have to paint it; there are very few if any ready-made and painted trenches available. Given the size of some of the scenery items you will be making, it isn't always practical, or economic, to use normal modelling paints all the time so I have turned to the DIY store and the range of mixed paints that are available.

I tend to use the Dulux mixed paint range, with a deep brown and a muddy brown colour in the Rich Praline shades being the most useful. There is a shade in the same series that is much lighter, and I did once use that as a top dry-brush but it gave the effect of light snow, which wasn't exactly the look I was after! Caramel is also a very useful colour for woodwork, as is silver-grey which you can use to represent aged woodwork. For the concrete bunker structures I use different shades of grey.

There are some items for which you can still use normal modelling paint. One of these is sandbags, for which I use Vallejo Desert Yellow and Iraqi Sand, and another is corrugated iron, for which I use Army Painter Rough Iron. For small areas of woodwork Vallejo Old Wood is also very useful.

Spray Primer

There are a number of ranges of spray primer available, but I tend to buy ours at one of the many discount shops that are around. You will need quite a bit of it, so best to buy it as economically as you can.

For overall base painting I use a textured wall paint which goes by a range of names, the most well-known one being Artex. This gives a very durable but slightly rough surface which nicely represents churned-up ground. There are other basing products around such as Basetex or something similar, but given the quantity I will need that may prove to be too expensive.

Sand

Many modellers use dried sharp sand as a basing material for their model figures and you can also use this for your terrain boards if you want to. It's especially effective for representing the sort of dry terrain that you might find in the Middle East or Gallipoli.

Flock

If you are going to represent an area of the trenches where it is not muddy and the ground has not yet been churned up by shellfire, you will need plenty of flock or static grass to represent the grassed areas.

Materials

Brushes

You will need a range of brushes, both for painting and for applying adhesive such as PVA. For doing the main boards you can use fairly big brushes, but when it comes to doing the work on the trenches themselves you'll find that smaller long-handled artists' brushes are better, provided the bristles are robust enough to do the job.

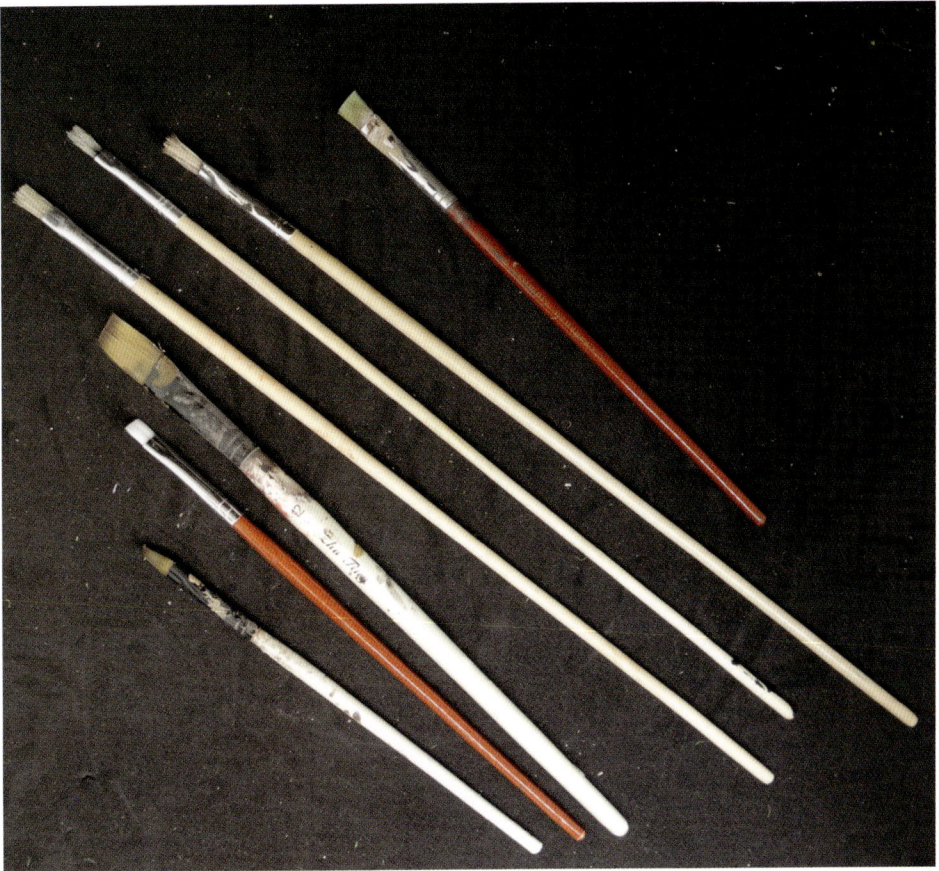

Cotton

One of the features of communication trenches in particular were the festoons of telephone cables that ran along the sides of the trenches. A great way to represent these if you want to is to use lengths of cotton glued to the sides of your trench.

Mask

Cutting and shaping Styrofoam in particular can produce a lot of very small particles. They will go everywhere, and you will be finding evidence of your work in the modelling room for years to come at unexpected times! To be safe it is always best to wear a good mask when you are doing this work.

Chapter 3

COMMERCIAL OR SCRATCH-BUILT?

One of the first things you will need to decide when building a trench system is whether you want to have a ready-made one or to build your own trench system yourself. Both have advantages and disadvantages which I will discuss, but ultimately this comes down to a question of personal choice and what suits your available resources in both time and money.

With the growing interest in the First World War, terrain manufacturers have been producing an increasing number of trench systems in recent years, in almost every scale. Not every range is comprehensive, nor do many readily mix with each other, but you should be able to find enough in your chosen scale if you want to go that way. This ready availability is one of the main advantages of commercially-made terrain, as in most cases you can buy some pieces and then gradually add more as you want to within your budget. This allows you to slowly build up your trench system until you have enough to stretch across your table.

As you can buy individual pieces, this also means that commercially available trench systems are very flexible. You can set them out in any number of ways, which will allow you to have an endless variety of games over different trench set-ups. As we will see later, some model trench systems also have differences between trenches, so that some appear to be more for a British force and others more for a German one.

Further factors to think about, which will no doubt be important to many of you, are things like transport and storage. As most commercially available trench systems come in individual sections they are much easier to store in a box and so don't take up a great deal of space. This is particularly important to bear in mind when space is at a premium and you need most of it for your model soldiers! These trench systems are also easier to transport to and from the wargames club for a game, and the ability to quickly set up and then break down a trench game will be a very important element that you'll need to think about carefully when considering introducing trench systems to your gaming.

Given the cost of some trench sections, there will inevitably be a temptation to think 'I can make that cheaper myself.' This may well be true, but it brings us to another advantage of pre-made systems which is the saving in time. Finding the time to paint and model is often as much a barrier to completing a wargames project as money, sometimes even more so. If your time is very limited and can more productively be spent on painting figures, it is less likely that you will want to spend much more of this valuable asset on scratch-building and painting trench systems. Having a ready-made system that all you have to do is paint, and even that perhaps in stages, can be a massive saving in the amount of time you have to spend on building your wargames terrain.

Also don't forget that you don't need to stick to the way in which the models for commercially available trenches are presented. Although most are fine as they are, many of them can also be improved upon with a little work and I will show you how to do this in a later chapter.

One of the main disadvantages of commercially available terrain sections, however, is cost. As each piece is individually available, making up a trench system can prove to be quite expensive, and this is all the more so if you want to make two trench systems for opposing sides. As the scales increase in size, naturally so does the cost in relative proportions and although in terms of space you will probably need fewer 28mm trench sections to reach across the table than you would if you were doing the same game in 15mm, the cost will almost certainly be comparatively more.

It also very much depends on the size of the range being produced by manufacturers. One of the main features of a firing trench was the division of the trench into firing bays by traverses, where the trench zigzagged, usually at right angles, to keep the blast of shellfire to a minimum. Some manufacturers have these traverses as a feature, but others will require you to buy four 90-degree turns to enable you to represent this. As the usual trench layout had a traverse every few yards, you can quickly see how the cost of doing this would rapidly spiral. Not only that, but the inclusion of such turns doesn't look very realistic on the tabletop.

This brings me to one of the other, rather more obvious, down sides of ready-made terrain, which is that by definition it has to sit on top of your wargames table. Unless you are gaming in the flooded terrain of Flanders where the trenches were often built up above ground level, the vast majority of trench systems across the Western Front and elsewhere were below ground. As ready-made trenches cannot be dug into your wargames board, they don't quite look right somehow, especially if you then have a flat no man's land (except for shell craters that also rise above the surface) rising to another trench system on the other side. It certainly spoils things like line of sight for troops on the other side of a trench when you have to pretend that the trench is flat even though it is clearly above the surface of the table.

With all this in mind, you may decide that a scratch-built trench system is for you after all. Unlike commercially available trench systems, scratch-building trench boards will certainly give you the look and feel of trenches that are dug into the ground and you can build any size and layout that you wish. You can, if you choose to do so, build trench systems for both sides and a no man's land with shell craters and barbed wire.

Building trench boards will allow you to build accurate and complex trench systems, perhaps with the complete three lines of trenches (although slightly condensed in scale) leading back to command posts and aid sections, or perhaps even back to areas behind the lines and gun emplacements. You are, in effect, only limited by the size of your wargames table and the type of layout you want to depict.

The financial cost would probably be roughly the same as buying ready-made trench pieces as all you will be buying are the materials to make the trenches, but of course the cost in time will be very much more. Making each trench board does take quite a bit of time and you will also need the space to work on it. If you are fortunate enough to have a permanent games area this may be less of a problem, but if not then making trench boards can be extremely messy as we will see! If you are not doing this on your own but with friends, or even as a club project, that may go some way to helping with the cost, both in financial terms and time.

Storage and transport are also issues that have to be considered. Trench boards can take up a *lot* of space, and can sometimes be bulky and cumbersome to transport. If you have to carry them to and from club nights there will also be the risk of damaging them in transit, so they would need to be well-protected for the journey. However, on the plus side, once you have got them to the club it would only be a matter of laying them out on the table and you are ready to go, so there would be a saving in time from that.

For many the main advantage of trench boards is that quite simply they look the part and you can build them any way you want to. You are not limited by short or long sections of pre-made trench and trying to get them to go where you want them to, or by difficult angles or gaps where you just can't quite make the pieces fit. Your model soldiers will move and fight along trenches that look and feel right, and really have the sense of a trench game. Also, as we will see, if you build them so that each section of board is interchangeable, it will allow you a degree of flexibility in the games that you do so that you are not always attacking over the same piece of ground (although for some games this would be historically accurate!).

There is also an in-between option and we will look at this possibility as well. Even if you have decided to scratch-build your trench boards, you can always have individual pieces of terrain that sit on the top of the board. These could be strongpoints, perhaps from ruined buildings that sit either out in no man's land or just behind the front line and form a great objective for capturing or defending in a game. Or they could be shell-shattered woods that your trenches wind through, with machine-gun positions dug into shell-holes in the wood. Or even the old cliché: a crashed aircraft in no man's land as an objective. In the Projects chapter we will look more closely at building some of these to add an extra bit of First World War flavour to your game.

Chapter 4

BUILDING AND PAINTING COMMERCIAL SCENERY

In this chapter we are going to look at a number of different commercially available terrain pieces in the three most popular scales: 15mm, 20mm and 28mm. In each scale there are a variety of terrain items available from a number of manufacturers, but I have chosen to look in detail at one or two manufacturers for each scale.

There are, of course, a large number of different trench systems available, as a quick look online will tell you. In 15mm I have chosen to look at those in the Ironclad Miniatures range, but as well as Ironclad some trenches are produced by Javis and by Kallistra, which fit in with their hex-based terrain system. Battlefront also produce a limited range to go with their Flames of War models.

In 20mm both Ironclad and Javis feature again, but the most comprehensive range that I have been able to find is that produced by Early War Miniatures, which is the reason I have chosen to feature it here. Unlike other manufacturers the trench system produced by EWM is vac-formed, which has the advantage of making it much lighter than other trench systems that are made in resin but with the possible disadvantage that it may not be as robust in the longer term, though it would certainly survive being dropped from the wargames table better than most others!

For 28mm I have chosen the Amera range, which are also vac-formed and are in their Future range. They do also make a range of 'trenches', but they do not interconnect as well as the ones in the Future range. These are relatively basic but as we will see, with a bit of work they can be made to look pretty good. I have also included some of the Ironclad 28mm trenches so you can see how they compare to their smaller cousins in 15mm.

For each we will look at painting them in different ways at a wargames level and then how you could set about adding to or improving them. I'll show you how to do this in a step-by-step way so that hopefully you will find it easy to follow if you want to do the same. These are only guides, of course, and it may be that when doing it you find a different way that you are happy with.

A further option to bear in mind, although it is not one that I am going to go into in any detail in this book, is the growing availability of trenches that are 3D-printed. STL files for printing trenches are already commercially available online, and can either be the stand-alone kind that sit on your wargames table or ones that can be inserted into foam boards. As the files are digital they can be printed in any scale and seem to offer a great deal of flexibility. However, 3D printers are still not for everyone, and there may be down sides in terms of the amount of filament you have to use to produce each model and the time each one takes to print, which of course has an effect on the cost. I have no doubt that this will be the way things will go in the future, and some wargamers have already taken the step of buying a printer and producing their own models.

15mm: Ironclad Miniatures

The 15mm range of trench pieces produced by Ironclad Miniatures, who are based in South Wales in the UK, is pretty extensive. The trench sections are readily available from their website online, and Ironclad do attend many of the wargames shows up and down the UK. As with any manufacturer, if you want a sizeable amount of terrain from them at a show I would strongly recommend ordering in advance to give them the opportunity to produce it first!

Each trench piece is made from resin so is quite strong but can be fragile if dropped, and is well-produced with no distortion or bubbles. Making the pieces out of resin does allow for a good amount of fine detail to show and makes painting them a bit easier. Each trench piece has a trench width of 30mm, which is sufficient to allow for bases of that depth – for example, for Flames of War units – to stand in the trench. The trenches mostly have a single line of sandbags along the parapet and wooden planking on the trench sides. However, given that they are only 30mm wide, they are largely symbolic and have no room for a fire-step, for example, though in 15mm this might not be too much of a problem as most games in this scale will use multiple-based figures which will sit quite well in the trench.

The trench pieces available are as follows:

A long straight, which is 120mm long. They also do a short straight which is 60mm long.

T-junction (facing the rear)

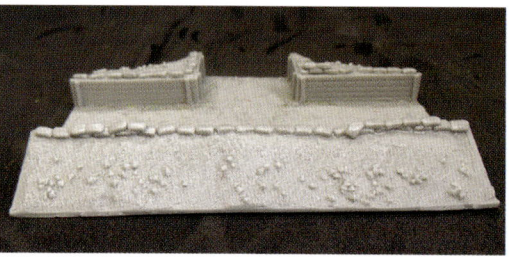

End section

Entrance section

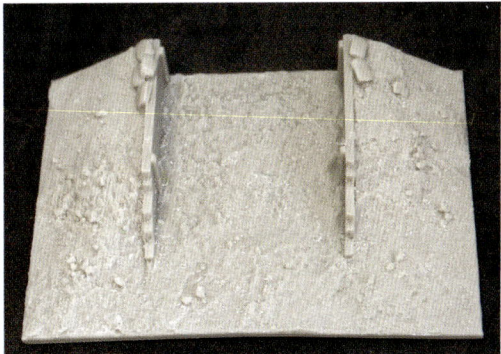

A 90-degree turn

A 45-degree turn

Z-sections

Concrete bunker

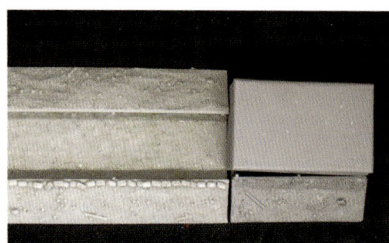
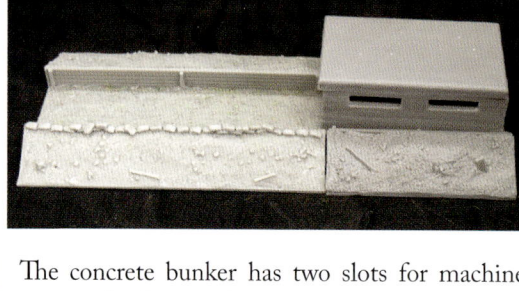

The concrete bunker has two slots for machine guns and a door in the back, and has a removable roof so that you can place figures into it. It sits on a resin base that does not reach the full depth of the model and only has a slope on the front side. Given the location of the door, you will probably want to use a short straight to lead up the bunker, and as an alternative you could use a bunker that the same company makes in 20mm. This is rather bigger in size, though still acceptable for 15mm, and can accommodate a based machine-gun team.

Building and Painting Commercial Scenery 31

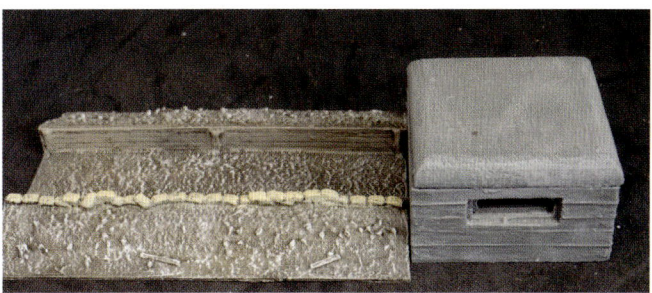

The range also has artillery positions and a log bunker. The chapter on Projects shows one of these emplacements, but the log bunker is really much more suitable for other periods or theatres so it has not been included here.

As these are resin sections, they are pretty much ready to paint as they stand. You can, if you wish, make some improvements to them and I will talk about those in a moment, but first we will look at painting them as they are.

Before painting them the first thing you will want to do is to prime them so that they take the paint properly. I do this quickly by using spray primer. Make sure you give them a good coat all over, and of course as you are using a spray paint do make sure that you do this in a well-ventilated area – I spray mine outside if possible – and please wear a mask. You can do quite a few in one sitting so that you are ready to go when they are dry. I have made myself a spray booth out of a cardboard box which seems to work well, and usually put the pieces that I am spraying onto a board of some kind. This means I can handle them easily, and I know some people who have made their own turntable for spraying. Once they have been sprayed I transfer them onto some newspaper to dry, as in the past I have found that if I didn't do this they stuck slightly to the board once they have dried, which made them tricky to remove. Transferring them can be a bit messy, and I have now taken to wearing disposable gloves for this part as I don't want to get my hands covered in spray paint!

Once the primer is dry, the next stage is to give each piece a basecoat of paint. Again it is useful to keep them on the newspaper while they dry, and you might find that the disposable gloves come in handy here too! I give each piece a good coating of dark brown emulsion all over and then allow them time to dry.

WWI TRENCH SYSTEMS

As the resin pieces have good detail and texture on the ground areas, one way to bring this out is to dry-brush them. For those who are unfamiliar with this technique, it is simply a way to bring out the raised detail on the model while retaining most of the base colour. Just put a little paint on your brush and then brush it on a surface (I use kitchen towel) to remove most of the paint. Once you have done that, gently brush it over the surface of the model so that it just touches all the raised detail. Try not to do this too heavily; it needs to be put on lightly so that it just covers the raised detail and doesn't cover the base colour too much.

Once the basecoat has dried, dry-brush the ground areas in front of and behind the trench and the middle of the trench with the muddy brown paint. This will create quite a nice effect on the model. You need to be careful when you are dry-brushing the middle not to get too much of the paint onto the sides of the trench where it is planked, though if you do you can easily just touch this up with dark brown so that it hides your errors!

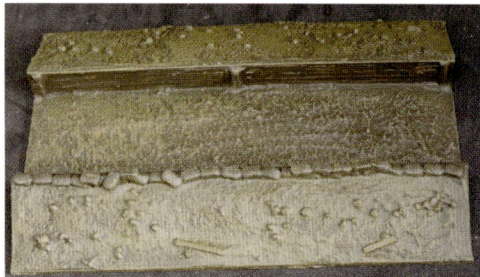 

If you wanted to, you could leave the ground this colour if you wanted to represent a really muddy area, but if you give the ground a very light dry-brush of a cream colour you will find that it brings out so much more of the fine detail on the model.

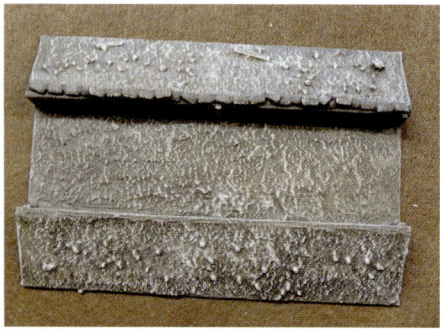 

Once this has dried, you can then go on to dry-brush the planked sides of the trench with a wood colour or with silver-grey. You do need to be careful here not to get the dry-brushed colour onto the other parts of the model, and you will need to use a much smaller brush for this.

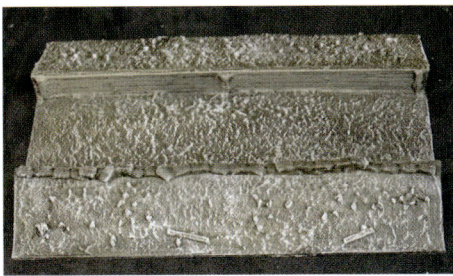

At this point you could finish the model off by painting in the sandbags. I do this by first giving them another coat of dark brown to make sure they are tidied up after the other parts of the model have been painted.

I then heavily dry-brush them with Vallejo Desert Yellow to give them the look of sandbags.

Finally, I complete the look with a further, lighter dry-brush of Vallejo Iraqi Sand.

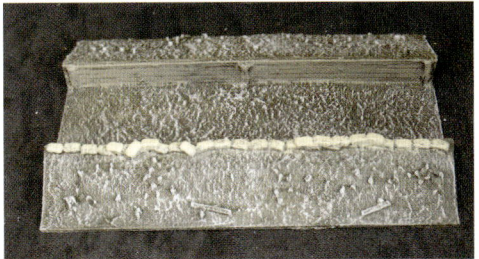

To protect the finished model, you could if you wished give it a coat of spray matt varnish.

All of these have just been painted and dry-brushed to give the effect we want, but there is another way of doing it. Once you have base-coated the model in dark brown, you could add sharp sand and static grass instead of dry-brushing with muddy brown and then highlighting.

To do this, simply cover the slopes of the trench model with PVA and then sprinkle dry sharp sand over it. I always find that it is best to do this over a sheet of newspaper so that you can catch any sand that spills from the model. Once you have done this, lightly tap the model (over the newspaper!) to remove any sand that has not stuck and set aside to dry thoroughly.

I tend only to do this on the outside slopes of the model. If you were to do it on the bottom of the trench it may not look as effective and could possibly get in the way of placing your based figures in the trench.

Once it has dried, you can add some static grass. Again, cover the areas that you want to have grass on them with PVA and then sprinkle the grass in place, again shaking off the excess over newspaper. You will almost certainly want to make sure that the bottom edges on both sides are grassed, and to put some random patches on the trench slopes themselves.

Given the size and level of detail that already exists on the models, the options for adding further adaptations to it are a bit limited. One thing that you could do, though, is to add more sandbags to give the trench the look of something that is more heavily built-up. At the moment the straight models only come with a single line of sandbags representing the parapet and nothing on the back or parados, though other models do have a line of sandbags on both sides. To add to the model, you could make a very thin line of sandbags using air-dried clay (see the 28mm section on how to do this) and then add them to the model, using something like UHU Solvent-Free glue as you are sticking them onto resin. You could add them to both the parapet and the parados if you wished to.

Be warned, however, that this is extremely fiddly as the sandbags are very small in 15mm and you may find that in fact it is a lot more trouble than it is worth.

Building and Painting Commercial Scenery

You could also add some to the T-Junction model to deepen the number of sandbags. By adding an end piece, or a short straight and then an end piece, you could represent a sap going out into no man's land or an MG nest out in no man's land.

One of the other features of the Ironclad 15mm trench system is that it could, relatively easily, also be used for 10mm figures as it stands.

In fact, in some ways the trenches suit 10mm figures better as they are nearer to the right depth. Of course the trench will be much too wide, but this might mean that you could put in a fire-step if you wanted to. It also depends on how your figures are based of course; if they are on deeper multiple-figure bases this may not be too much of a problem.

One of the flexibilities of this system is that you can use the pieces over and over to make any combination of trenches that you wish. Each piece fits seamlessly with the next and can therefore be used in any number of ways. However, setting up a trench system piece by piece will take time, and if you wanted to limit this but still retain the flexibility, one way in which you could do that is to mount some of the trench sections onto a piece of MDF board. In this way you can keep some of the flexibility of the system but make setting out the terrain just a little bit faster. The section on building trenches for 28mm Amera covers the principles of this; just apply those to your 15mm trench systems to get the same effect.

Try to make each board no more than about 30cm long as you may find that if you make it much longer the weight of the resin pieces will warp it over time. You can keep the flexibility by having T-junctions facing forwards in a few places so that you can add end pieces as saps, and facing backwards so that you can add a combination of short straights and 45-degree turns to make communication trenches. These could then connect with a second line of trenches, using the forward-facing T-junctions to join onto the communication trenches. Alternatively you could have a short straight and a 90-degree turn from a rear-facing T-junction leading into a concrete bunker position.

20mm: Early War Miniatures

In 20mm Early War Miniatures have in recent years produced an excellent range of trenches with some innovative ideas. Unlike other trench systems, each model comes with a couple of fire-bays separated by a traverse, which very neatly avoids the problem of a profusion of 90-degree turns and having to buy a lot of models to get the effect you want. Early War Miniatures products are available online, and they attend a lot of wargames shows in the UK and in Europe.

Where others have made their trenches in resin, these are made from vac-formed plastic. This means that they are very light and almost unbreakable, and each trench model will stack on top of another similar model, so making storage a bit easier. An innovative aspect that I had not seen elsewhere is to join the trench pieces up using small rare earth magnets, which holds the system together and prevents it moving about when on the tabletop. However, the down sides of vac-formed plastic are that it can sometimes go out of shape very slightly so that the edges appear raised on the table, and to a certain extent the level of detail that you can see on the model is slightly less than it would have been had it been made from resin, though of course that would have massively increased the weight and cost!

One of the things that EWM have done is to make different trenches for the British and the Germans, though of course the German trenches could also be used for the French. As you will see from the examples below there are slightly fewer trenches for the Germans and they are not as flexible to use as the British ones, but they are there nevertheless and there are things you can do with them. Each trench piece is designed to link with another piece, but unfortunately the 'British' and 'German' trenches are different widths and depths so aren't fully compatible with each other. All is not necessarily lost, however, and although it means accepting some visual compromises that are not ideal, you can use them together as we will see below.

For the 'British' style, the trench pieces have mostly planked sides with areas where the sides are reinforced with wrinkly tin. Each piece is roughly 30cm wide by 25cm deep. The firing bays have a fire-step which will take a 20mm figure on an individual base, though it would not be suitable for multiple-based figures. The bottom of the trench is also wide enough to take an individual 20mm figure base.

The items available are as follows:

British trench with two firing bays and shelters.

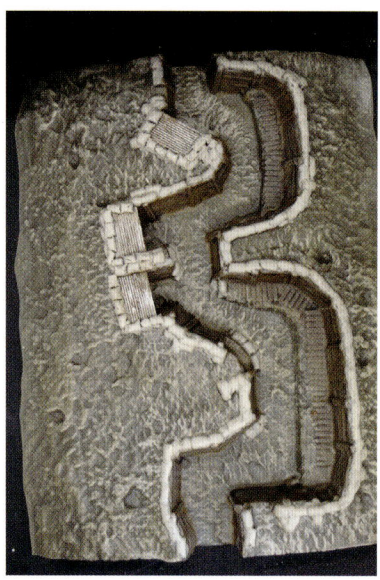

This has small shelters, though a variant has a mortar position instead. This would most likely form the basis of your trench system.

British trench with two firing bays and rear exit.

This is a trench with an exit to the rear that would join up with a communication trench to lead to a second or even third line if your table is big enough!

British communications trench.

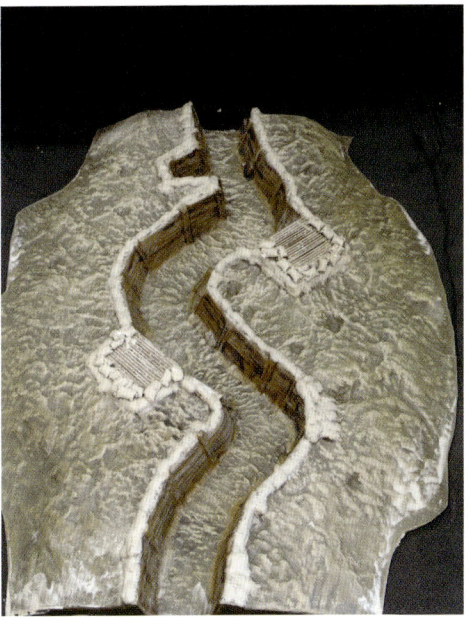

A very handy piece of kit and one of the few models I am aware of that accurately represents a communication trench. This one has an area to represent the latrines (very useful for your model soldiers!) and shelters or storage areas.

British trench with two firing bays and front exit.

Another useful item, that will allow you to represent a second line with the communication trench joined on to it, or to join to areas where the troops have sapped forwards.

There is also a Forward Sap; one of the few items made that represents a sap going forward from the main trenchline towards the enemy. The sap is shown as if it was unfinished, with two sections at the front that could be used as defensive positions for riflemen or machine-gunners.

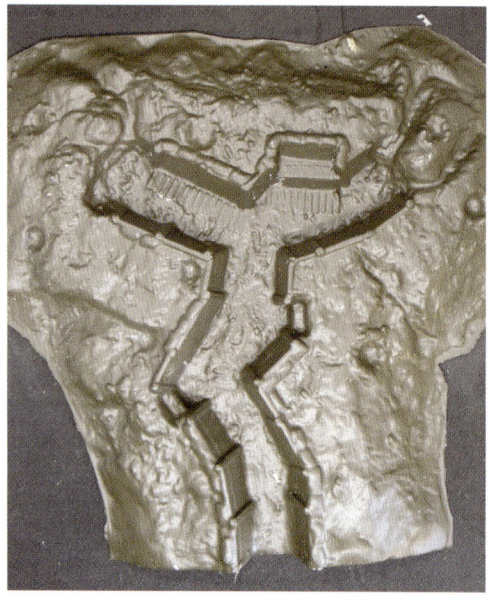

Building and Painting Commercial Scenery

A 90-degree turn.

This enables you to turn your trench system through 90 degrees.

A 45-degree turn.

Also usefully you can get a 45-degree turn, which can go with the other turn to make a trench either turn in or turn out.

As these models are vac-formed, they come on a plastic sheet and the section on Amera Models in 28mm has some more detail on how to prepare this.

Once the model has been cut out it's time to paint it, and for this we take much the same approach as we did with the 15mm resin model. As these models are vac-formed plastic there is no real need to give them a coat of primer, though of course you can do so if you wish. I normally start by giving the model a coat all over of dark brown emulsion to act as a base.

Once this is done, a simple dry-brush of muddy brown over the outer areas and inside the trench where the walkways and fire-steps are will give it a muddy, earthy appearance.

You can now build up the dry-brushing with a further, lighter layer of caramel and then perhaps even the cream very lightly to pick up the raised detail, though as these are vac-forms it won't be as sharp as it would have been if it had been made from resin. Yet the detail is there, and you can bring it out with some careful dry-brushing.

Next you can quite heavily dry-brush the planked areas with a wood colour, or a silver-grey colour if you wish. If you have gone over them too much when you were dry-brushing the muddy colour, just re-touch those areas with the base colour and then dry-brush them when they have thoroughly dried.

Many of the models come with wrinkly tin to cover the shelters or dug-outs at the back. You can glue these in place as they are, using something like UHU as it won't melt the vac-formed plastic. Once they are in place, you can add a row of sandbags around the area of the shelter as this adds to the realism and helpfully covers any gaps! Make the sandbags using air-dried clay (using the technique shown in the 28mm section) and glue them in place. Once they have dried, you can paint them Vallejo Desert Yellow and then dry-brush them with Vallejo Iraqi Sand.

As well as sections of wrinkly tin, the EWM trench systems also come with a variety of trench stores, signposts and ladders. You can also buy these separately. They are made from white metal and so take paint very well, and you can use them to decorate your trenches and give them that 'lived-in' look.

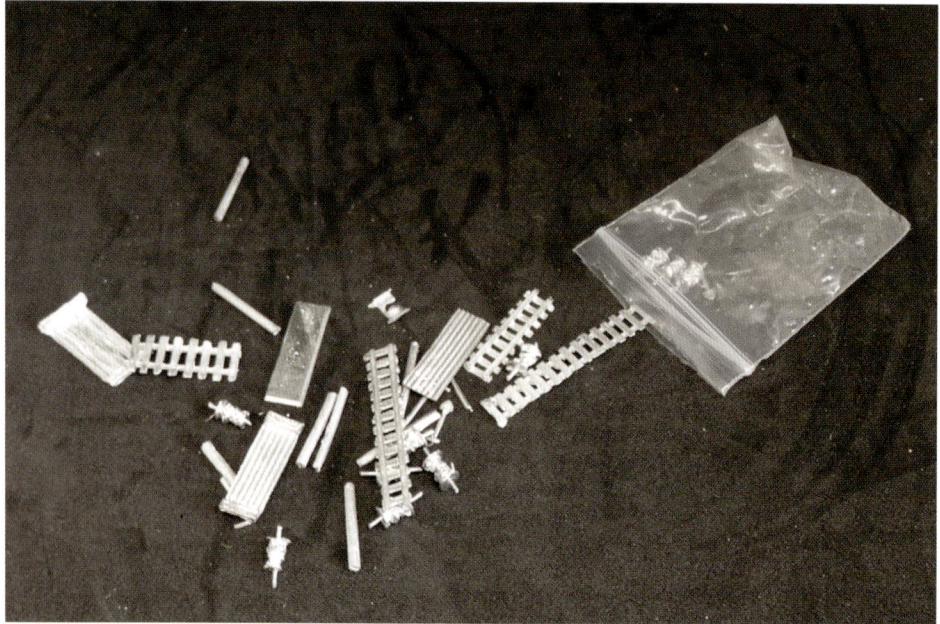

Building and Painting Commercial Scenery 41

WWI TRENCH SYSTEMS

One of the most useful accessories that EWM make in this range is a small steel plate, representing the kind that was often used by both sides as protection for a sentry or sniper's position. This really adds atmosphere to a trench model, and you can easily slot it in between two sandbags made from air-dried clay.

Lastly, if you want to do this, you can glue the rare earth magnets inside each side of the model (making sure you get the polarity right!) so that it will attach to the other models in the trench system. You will need four magnets for a normal trench section, but of course if it has a front or rear exit you will need six, so the number that you will use soon adds up. However, it does work, and holds the trench system together very well.

The German trenches are described by EWM as 'linked' trenches, though both systems link equally well. They are slightly larger than the British ones, being about 40cm wide by just over 27cm deep, which is where the matching-up problem with the British trenches occurs as they are only around 25cm deep.

Unlike the British trenches, one of the difficulties with the German trenches is that all the models have an exit to the rear, and unfortunately this exit tapers down as it reaches the edge of the model, so it is very difficult to join a communication trench onto it. This is most likely because I think this was the first range that EWM made, and later lessons enabled them to make the British ones more compatible with each other.

All the German trenches have wattled sides with some areas of wrinkly tin, such as the supports for the fire-steps. Like the British trenches, the German ones have a fire-step in each fire-bay that is wide enough for an individually-based 20mm figure and the trench behind the fire-step will also take based figures.

The models available are as follows:

German trench model with two firing bays.

This is very similar in style to the British trenches in that it has two firing bays separated by a traverse, but this model also has a small position that could be used for a machine gun or a sniper. Instead of a shelter at the back it has two areas that represent dugouts.

There is a trench that is very similar to this model, except that one of the positions at the back now also has a firing bay and fire-step, and the model does not have the smaller firing position that model 1 has. This could be used as a rear support position which was quite common in German trench systems. The other side at the rear has a dugout with an entrance facing the front.

German trench deluxe model with one dugout.

Building and Painting Commercial Scenery

This model is identical to model 2 above, except that the dugout now has a removable resin roof so that figures can be placed inside.

German trench deluxe model with two dugouts.

This model now has a dugout on each side with a resin roof. As a bonus, there is an additional resin item that you can purchase separately that represents a bunker. This fits over one of the dugouts in this model and is a handy way to represent either a machine-gun bunker or a *Stollen* shelter.

There are also 45-degree turns available so that you can turn your trenchline at an angle.

As well as British and German trenches, EWM also make some very useful stand-alone strongpoints. These are a defensive position that could, for example, be part of a third line or a redoubt or 'keep' behind the front lines. They do not link up with any of the other trenches but are designed to be individual pieces.

If you are content with having an exit at the back of each trench section, you can prepare and paint the German trenches in exactly the same way that I have described for the British trenches above. If you want to limit the number of rear exits or to have a trench join with a communication trench, there are a couple of ways you could do this.

One is to accept that the depths are different and just use a section of British trench next to a German one.

I have to admit that this isn't at all ideal from a visual aspect, but if you are happy enough with it then you can go that way. It does enable you to have communication trenches in the same way that you have for your British trenches, and to have perhaps second or third German defence lines. The Germans did use planking and wrinkly tin to support their trench sides, just not to the same extent that the British did, so historically it would be OK. You could perhaps paint the ends of the British communication trench with some of the textured paint so that when you do paint and dry-brush it, it will have some texture to it and not look too stark. Alternatively, you could just use the British trench sections for both sides.

If you don't like the fact that every German trench section has an exit to it, one of the ways to overcome this is to add a small piece of Styrofoam, or even perhaps a chunk of balsa wood, to the area where the exit is to create a plug. This would then mean that you would have to cover the entire outer area of the model with textured paint so that it would fill the gaps and cover the area where the 'plug' is, but it can easily be done. If you were content to simply use one German trenchline, this may be a way forward so that only some of the trench models have exits to the rear. Once you have 'plugged the gap' and covered the model with textured paint, you can paint and dry-brush as normal.

Building and Painting Commercial Scenery

EWM also make a useful range of shell-holes. Again these come on a vac-formed sheet and need careful cutting out, but they do give you a number of shell-holes to use in no man's land and around your trench system.

28mm: Ironclad, Amera

For 28mm figures I am going to look at two very different options, these being those made by Ironclad again and some models made by a company called Amera Mouldings.

Ironclad, as I mentioned above, are a company based in South Wales who attend a lot of the wargames shows in the UK. The models in their 28mm range of trenches are very detailed due to their larger scale, and there are options in this range that are not included in the smaller ranges. Made from resin as with the other scales, these are weighty pieces that will sit very well on the wargames table. The trenches themselves are about 30mm wide, and both sides are planked so there are no options to differentiate between the trench styles of the different nations. The trenches are also about 20mm deep, so although you can easily stand individually-based figures in them, there isn't the width or the depth to allow you to build in a fire-step, for example. Additionally, the trenches only really come to about chest-height for a 28mm figure.

This works for figures who are standing firing, as their rifle is just over the top of the trench. There are only a few sandbags on the sides of the trenches, so if you wanted to make them look a bit more realistic you could add some more sandbags using air-dried clay as described elsewhere in the book.

The following models are available:

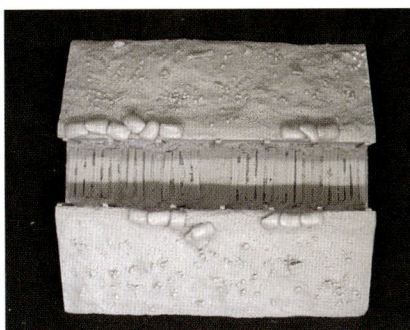
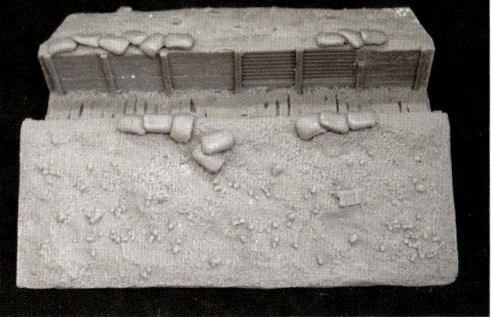

A long straight, which is 150mm in length. This length of trench is also available with a shell-hole crater in the middle of it.

There are also short straights, which are only 75mm long, and as well as the usual trench sides these also come with the options of having an entrance to a dugout or 'command post' in one side of the trench or a section where the trench side has been repaired with sandbags.

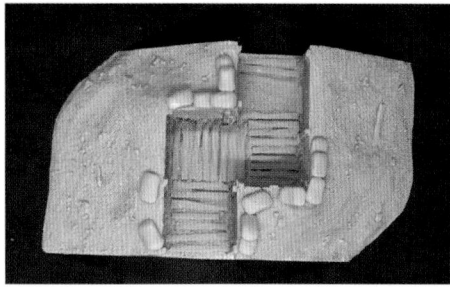
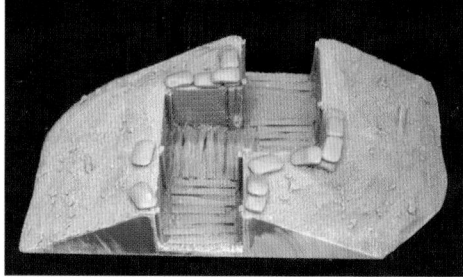

There are two 'zigzag' sections, which as with the 15mm models enable you to put in traverses between fire bays.

There are also T-junctions, end sections and entrance pieces, all of which are very similar to the 15mm versions, just in a larger scale, so I have not included pictures of them here, as well as gun emplacements.

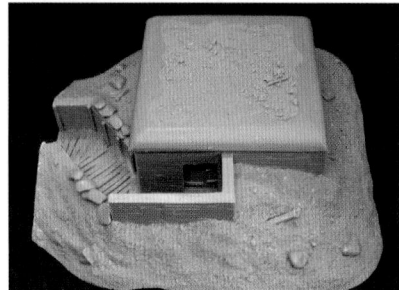

Building and Painting Commercial Scenery

One of the really impressive items in the Ironclad range is this machine-gun bunker with an entrance. It just really looks the part and would make a great objective for a trench raid.

If you wanted to paint these up quickly to get them onto the wargames table, then the process described above for their 15mm cousins will work just as well. A good spray primer, followed by a basecoat, and then paint as normal.

Once you have done the terrain as normal, the bunker should first be painted a dark grey

and then dry-brushed with mid-grey

WWI TRENCH SYSTEMS

and finally a lighter grey.

The inside of the bunker could be kept just the dark grey so that it appears to be darker.

Building and Painting Commercial Scenery

WWI TRENCH SYSTEMS

As with other models, you can add more sandbags using the air-dried clay technique, which I will describe in the section on detailing the Amera models.

Amera Plastic

Amera Plastic Moulding make a wide range of models in vac-formed plastic. They are available from Amera online, though I don't think they attend any wargames shows.

They do make some 'trenches' in their 1/72 range and although these do have an interlinked shell-hole section, the models are pretty basic and don't really represent a detailed trench system.

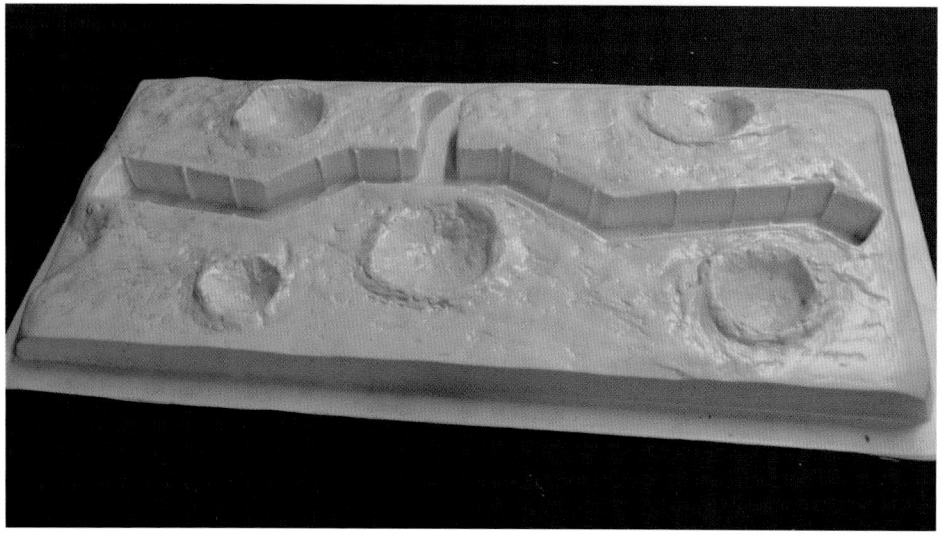

Building and Painting Commercial Scenery

They also make a short trench in this range, which is more representative of a small strongpoint than a trench.

The trench pieces that are most useful for our needs are those in their Future range. These are trenches that are designed perhaps with sci-fi games in mind but will fit our needs perfectly for First World War gaming in 28mm.

The following models are available in the Future range:

A straight section of trench. This is 30cm long by 20cm wide and 3cm deep. There is a similar section that has some shell damage in the side, and a shorter section of 15cm length.

T-junction. This is also 30cm long and has an entry point halfway along it.

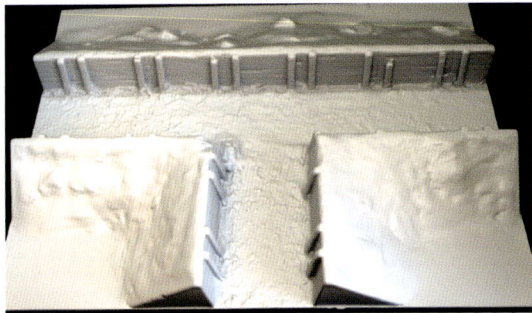

Trench end section; this is also 15cm long.

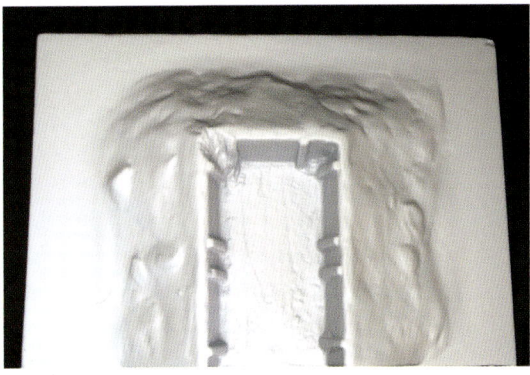

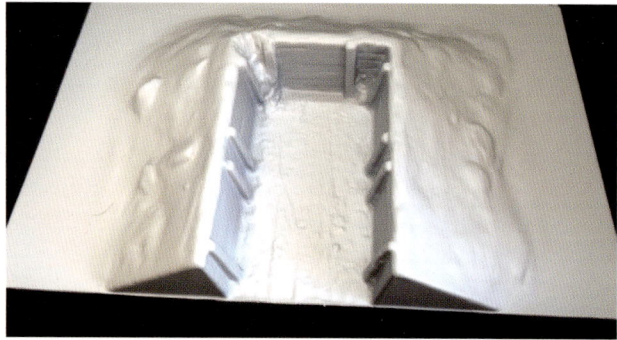

A 90-degree turn section.

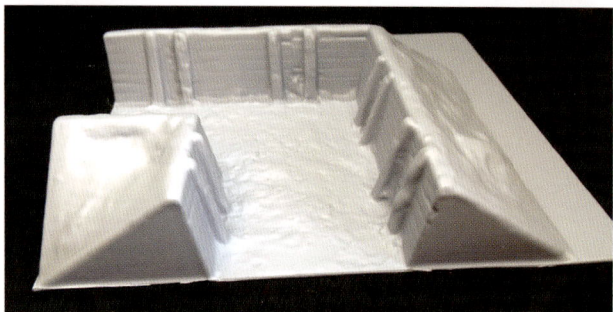

There is also a large machine-gun bunker, with a trench entrance that will match with, for example, the T-junction entrance.

The Amera models are relatively basic, with some planking slightly engraved on the trench sides and some support posts in place. However, one of the advantages of these pieces is that you can detail them very easily and I'll show you how. Some of the techniques discussed below will also come in very useful should you wish to scratch-build your own trench system, as we will see in the next chapter.

If you are going to base these models on MDF, they will first need to be cut out. In many ways this is the most important and delicate part of the operation, as if you cut into the actual model when you cut round the outside it will not sit squarely on the board when you glue it, so you must do this *very carefully*. Try to cut a little over the edge into the spare plastic area, as you will always be able to trim it a little later if you want to.

One way of cutting out vac-formed models is to draw a line with a felt-tip pen – preferably black – around the outside edge of the model itself. You can then carefully cut round that with a craft knife, using the line as a guide. It might be very tempting to try to cut it out with scissors, and I know that on occasions I have fallen for that temptation too! While it might seem a good way to save time, this can cause problems as no matter how careful you try to be, scissors rarely cut evenly and might leave you with a jagged edge. This then needs tidying up, and in doing so you can inadvertently cut into the model itself.

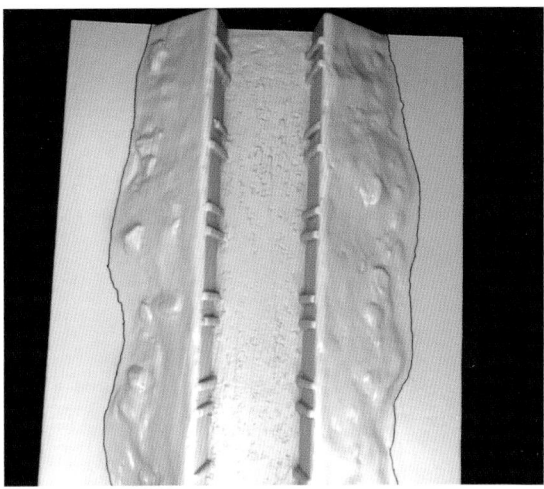

However, if you are not going to base the models or are content to leave them as they are, then it may be better not to cut round them as this will help them sit better on the wargames table and makes them more stable.

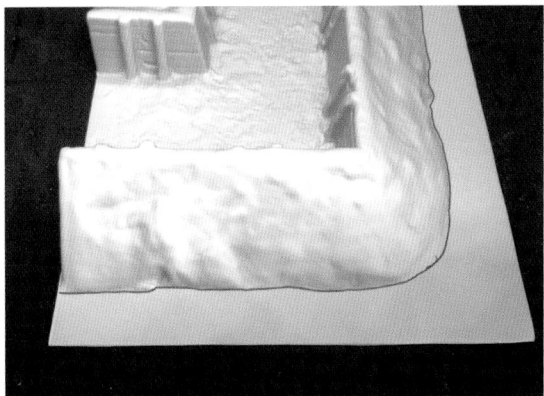

If you have decided to base these vac-formed pieces, you will need to cut a piece of MDF to size (remembering to wear a mask, of course). Before you glue the trench to the base you will need to seal it as it is MDF and you don't want it warping should moisture get into it. A watered-down mixture of PVA will achieve this, but another way is simply to paint the base on both sides and the edges with two coats of emulsion. I use our usual dark brown colour for base-coating for this, and then glue the plastic trench piece onto the base using UHU.

On the other hand, if you want to use the system of keeping the pieces together by using rare earth magnets as we mentioned for our 20mm models, then basing doesn't always work very well. Once you have glued the magnets on and then fixed the model to the base, you may well find that over time the magnets become detached and are then rolling around inside the based model, which is not what you want at all! A further disadvantage is that the based model has a rigid structure, which won't help when magnetizing the two pieces together as they need a certain amount of flexibility to work properly.

The first decision you will need to make is whether you want to build the trenches as broadly representative of British or German (or perhaps French) trenches. Let's start with some British trenches.

As we have mentioned, one of the distinctive factors of British trenches is the planking and wrinkly tin supports on the trench sides. To represent this, we are going to use sections of coffee-stirrers.

You will first need to decide whether you want to have a fire-step on one side of the trench. There is room for this, as the trench is 55mm wide internally so there is room to have a fire-step and still have space to put figures in. The trench models are 3cm deep so with 28mm figures you won't want to make the fire-step very high, perhaps only 3mm or so just to represent it. The best way to work out how high you want it is to put a 28mm figure in the trench and then see how high the figure stands above the trench sides on a 3mm fire-step. Ideally a rifleman firing from a standing position should be able to have his rifle over the top of the trench. You will probably want to make the fire-step about 20mm wide to allow for a figure to stand on it and balance properly.

You can do this either with a piece of 3mm deep foamboard, Styrofoam or balsa sheet. If you are using Styrofoam you will need to glue it in place using a solvent-free adhesive. If you use ordinary glue it will melt the Styrofoam, though you could use ordinary glue if you were using foamboard or balsa strip.

I've used balsa strip, so first had to cut out a piece

and then glue it in place.

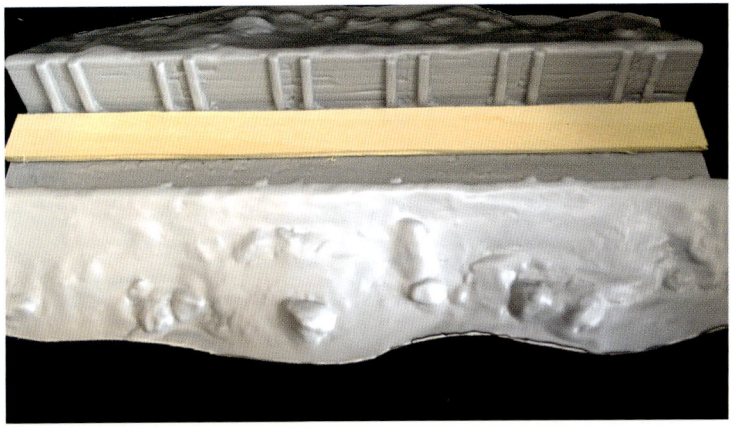

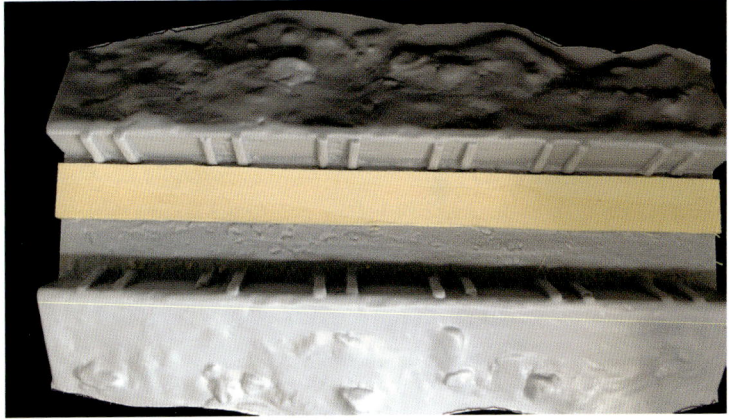

Once you have your fire-step in place, it's time to complete the trench sides. Measure roughly how long you'll need each section of coffee-stirrer to be to fit between the two supports on the trench model, and then cut the pieces you need to the right length. It's useful if you can cut a number of pieces rather than one at a time as this speeds up the process of sticking them on.

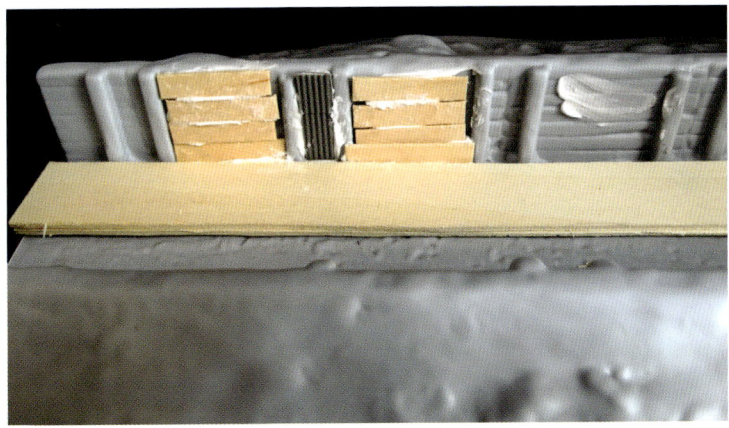

Once you have them, stick each piece onto the trench sides, which will give you a more detailed trench side. You can intersperse this with cut sections of plastic wrinkly tin to add variety. I've used solvent-free adhesive to glue these in place. It is incredibly messy, but a much quicker way of sticking the boards in place.

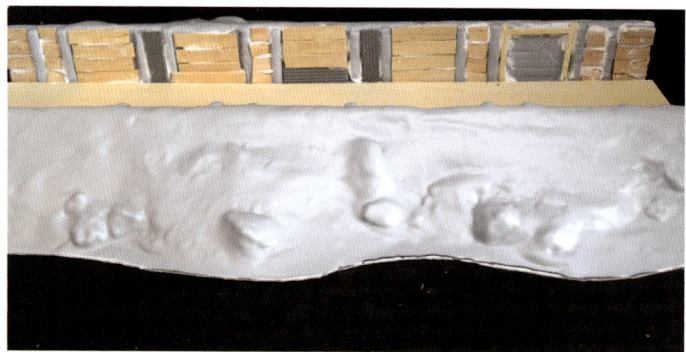

When you are adding your trench sides, think about where you want to have your dugout entrances. These would typically be on the side of the trench facing the enemy, so that any shells that did fall into the trench didn't fall straight into the dugout. It will also depend on whether you are modelling a front-line trench or one in the support or reserve lines, where command dugouts and aid posts would be more common. On average you might want to have perhaps one or two dugout entrances in a long straight section.

When you come to build these, just leave a gap in the planking with some matchsticks to represent the entrance, like this:

Building and Painting Commercial Scenery   61

You can then add a piece of tissue paper if you like, to represent the gas curtain that dugouts had across their entrances.

Once this has dried, paint in with Vallejo Canvas.

Once you have completed the trench sides, the next thing to complete is the trench floor. Again there were some differences between the British and German trenches, with both trench floors tending to run from side to side across the trench, but with the Germans sometimes having theirs running end to end. To represent this, you could add pieces of cut-up coffee-stirrer at intervals along the trench. You don't want them to be too close to each other, perhaps 5mm or so apart, and you'll need to think about aligning them carefully to go round the corners of the 90-degree trenches.

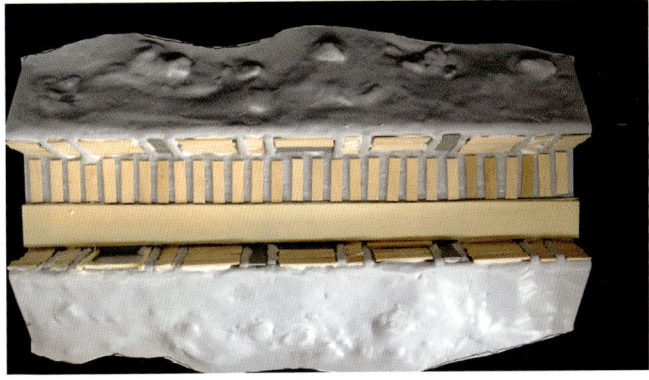

WWI TRENCH SYSTEMS

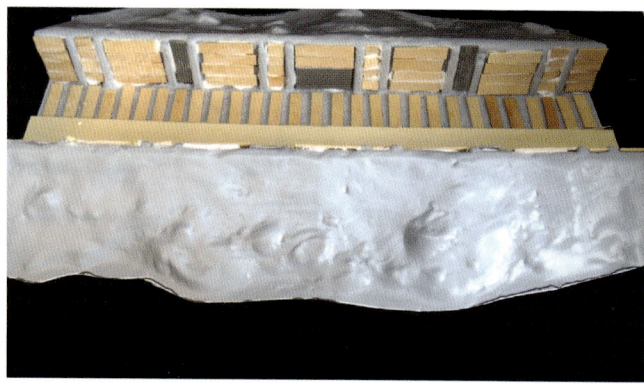

Once you have done all that, you can add the textured paint to the sides of the model to give it some texture when you come to paint it. You will need to let this dry well before painting.

One of the disadvantages of the Amera models is that, especially with the 30cm-long straight, they are one length of trench without any traverses. You could remedy that by only using short straights and then having four 90-degree turns to represent the traverses, but not only does this get quite expensive, as we have talked about in the chapter above it doesn't look very realistic.

One way to avoid this is to put small traverses into the straight model. On a 30cm straight, you could, for example, have two traverses in the section with fire bays in between. To achieve this, simply cut a section of Styrofoam about 5cm wide and the height of the trench.

You will need to angle the back of it by cutting carefully with a craft knife so that it will fit into position on your trench.

Building and Painting Commercial Scenery

You could then gently shape the top and sides a little to make it look less square, and then glue it in place with solvent-free adhesive.

You could also add a machine-gun position using a small block of Styrofoam or balsa, and then add some balsa to the sides and top.

Once all this has been done, you can then start to paint the model all over with the dark brown basecoat paint. One of the drawbacks that you will find with using textured paint is that no matter how thoroughly you think you have painted it, there will be small holes of white staring at you from every angle from which you look at it. This is especially so where you have had to use the textured paint to fill gaps, or perhaps put it on a little thicker in some places than others.

Just about the only way to resolve this is to wait until the basecoat has dried and then go round the model several times, looking at it from different angles as some of the 'white holes' will only be visible from certain angles. However, once you see them, they will annoy the life out of you until you have them all taken care of!

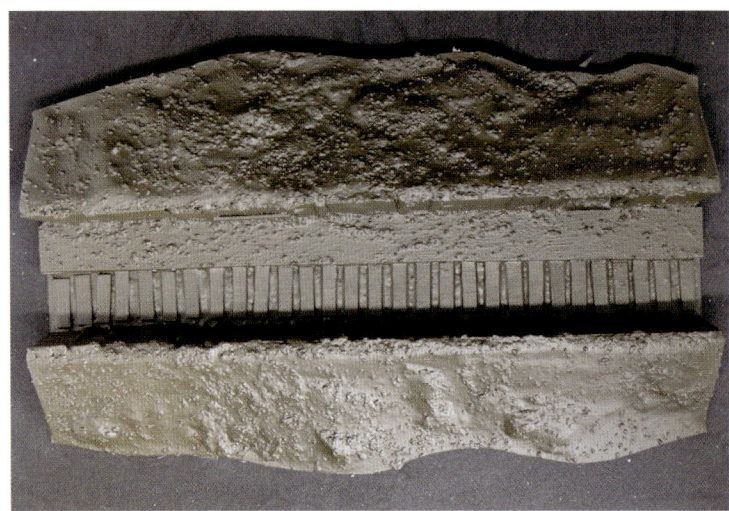

When you have finished dealing with all the white holes that you have been able to find, you can then paint the model in the way we have done for other models by dry-brushing first with the muddy colour and then with the lighter caramel or cream colour. You could then dry-brush the planking with a wooden or silver-grey colour, being careful when you do the planking on the floor surface not to get too much paint onto the ground. If you do, just a gentle touch-up with the muddy colour usually hides it.

Building and Painting Commercial Scenery

The next step is to make some sandbags. You will need these to run along the top of the parapet and parados of your trench, possibly two levels high in places or even higher, for example where you might have a machine-gun or sniper's position. Sandbags are also great for covering gaps and for generally piling around places such as the entrances to dugouts.

To make these, we are going to use air-dried clay. You should first tear off a chunk of clay, but as it lives up to its name and dries quite quickly in the open air, do remember to cover the packet again as soon as you have done that. I tend to keep mine in an airtight bag and then inside a box.

Once you have your chunk of clay, find a board of some kind on which to roll it out. Please bear in mind that this stuff will stain, so do *please* make sure that whatever you choose to use is suitable. I tend to use an old kitchen cutting board, but just find whatever works for you. You might also want to wear disposable gloves if you have them as it will stain your fingers as well!

You can then start gently rolling out the clay into a sausage-shape or several shapes. You want to make them reasonably even and not too thin; just judge what looks about right for 28mm by maybe putting the rolled-out clay near a 28mm figure.

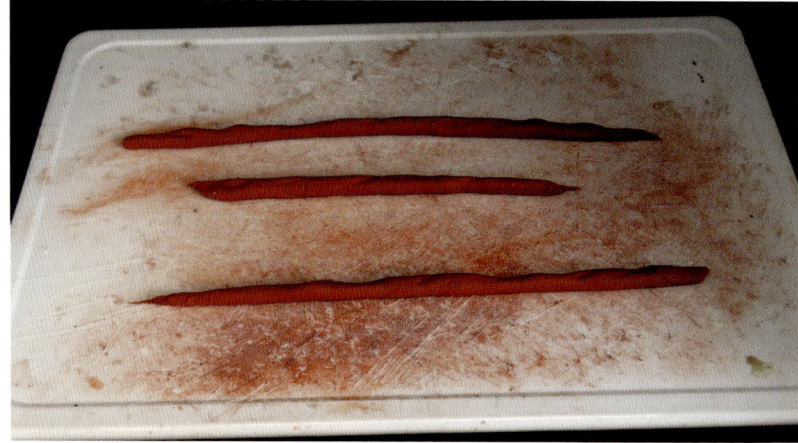

Once you are happy with the thickness of your roll, you can start to cut it up. If you do this with scissors, it will automatically leave a slightly indented mark at each end which makes the cut-up pieces look remarkably like sandbags.

Once you have your cut-up pieces, you can start to glue them down into the positions where you want them. I have used solvent-free adhesive, UHU and PVA for this and they all work reasonably well, although it depends what you are sticking them to, of course. For these pictures I have used PVA. Just run a little bit along the area where you want to put your sandbags, but not too far as you don't want to run out of sandbags! You can then glue the first layer in place.

You will need to work fairly quickly as the pieces will start to dry out. You will find that you can gently shape them as you put them in place so that they take on the appearance of sandbags.

Once you have the first layer done, you can then repeat the process to give you a second and even a third layer if you want to do so. For the second layer remember that sandbags don't sit neatly on top of each other; in fact, the less neat they are, the more realistic they look. Try to place them so that they are not sitting exactly square with each other, and again slightly squash them into place.

Once the sandbags are dry, paint them with the dark brown paint.

Next, dry-brush them with Vallejo Desert Yellow and then Iraqi Sand as before.

The wrinkly tin areas should be painted a gunmetal colour, and you could also run a rust-coloured wash over these areas if you want to. If you use Army Painter Rough Iron, this already has a slightly rusty shade to it.

Lastly, if you wanted to add some grass to the outside slopes, cover the areas you want to with PVA and sprinkle on some static grass.

Building and Painting Commercial Scenery

To make some German or French trenches, we are going to do most of the same things we did for the British trenches, except that where we had used planking we are instead going to use lengths of thin string. This works quite well to represent the wattled material often used on the sides of German and French trenches.

So instead of cutting sections of planking, you now cut lengths of string. The best way to apply this is to spread solvent-free adhesive onto the area you want to cover and then stick the lengths of string on.

You will need to allow the string time to dry, but once it has you can paint it in the normal way. Most wattling was made from thin strips of wood, so dry-brushing it the wooden or silver-grey colour has quite a good effect.

As we have mentioned before, the Germans did use planking and wrinkly tin too, but to perhaps a lesser extent than the British. This does mean that you can feel free to mix and match wattled string sides, planking and wrinkly tin as you wish to get a good effect.

WWI TRENCH SYSTEMS

WARGAMES TERRAIN & BUILDINGS

Building and Painting Commercial Scenery 71

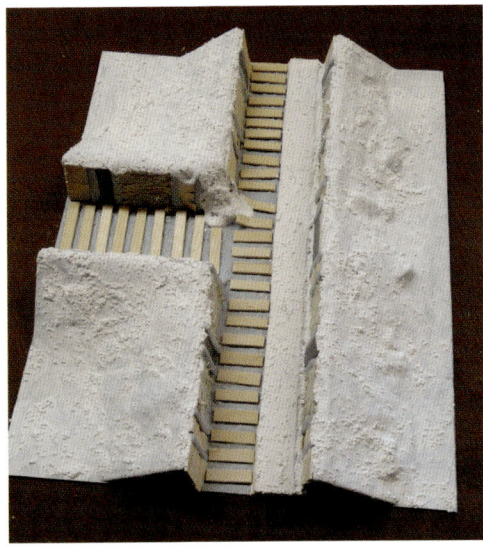

One of the features of this section of trench is a collapsed trench wall, which you can enhance using the textured paint.

WWI TRENCH SYSTEMS

Building and Painting Commercial Scenery

WWI TRENCH SYSTEMS

Chapter 5

SCRATCH-BUILDING

Having looked at some of the options for commercially available trench systems, let's now look at building one yourself. This is not perhaps the easiest option if you want a trench system for your First World War wargames as it takes quite a bit of work to build one, but if you decide that is the way you want to go, then I will show you how to set about building a bespoke trench system.

Much depends, of course, on the size of your wargames table. It's best if the tiles that you make can be interchangeable to some extent to give you a degree of variety in the table you set out and I will look at ways that you can achieve this.

Most of the boards in the following pictures were made for a bespoke trench system and are 120cm by 60cm. I got mine from a company in the UK called Panel Systems. They currently make a range of both blue and green modelling foam, with the blue coming in 120cm by 60cm and 60cm square, and the green just in 60cm squares. Blue has a greater variety of thicknesses, with the 60cm squares coming in thicknesses for our needs of 3mm, 10mm, 15mm, 20mm and 25mm, and the 120cm by 60cm coming in a thickness of 25mm. The green 60cm squares are in thicknesses of 25cm, 30cm and 40cm. Unfortunately you can't buy the squares individually from them; you have to buy a box, with each box having a varying number of tiles in it depending on the thickness.

Depending on the scale you are working in, you will most likely want to use the 20mm deep ones for 15mm, 25mm or 30mm deep for 20mm figures, and 40mm for 28mm figures.

The first thing that you will need to do is to plan how you want your trench systems to look. If you can, lay out all the boards on your wargames table or wherever you are planning to work on them, and start to mark out where you want your trenches to go. It is very important at this stage to think about the sort of layout you want to build. How many trenchlines are you going to have on each side? With 60cm by 60cm boards you can probably have up to two lines of trenches on a board, so if your wargames table is 6ft wide you could easily have boards with two lines of trenches on each side and a 60cm wide no man's land board in between.

However, if your wargames table is only 4 or 5ft wide, then you will have to think differently. Having two lines of trenches on each board could leave you with very little room for a no man's land, and although in some cases the trenches were very close to each other (especially in places such as Gallipoli, for example), that doesn't make for a very good game. If you can't build a board for no man's land to go in between your trenchlines, then you may have to accept having only one trenchline on each board with perhaps 20cm or 30cm of no man's land on each board. This will still allow for a good game while keeping the impression of two opposing trench systems. Alternatively you could vary it by having two lines on one

Scratch-Building

side and only one on the other; the choice is yours! That is one of the great advantages of building your own trench system, of course; what you build is entirely up to you to decide.

You should mark out your prospective trench system on the board with a marker pen. The key things to remember are to keep your lines reasonably straight, so using a straight-edge helps, and to make sure that the edges of each trenchline match up on each board that you do. In this way you can be sure that the boards will fit together in any combination you want. One way to make sure of this, after you have laid them all out, is to swap the positions of each board around a few times to make sure that the edges you have drawn are all lined up. Don't forget that trenches were not straight: they had fire bays which could perhaps be one or two multi-figure bases, or maybe four or five individual bases wide and then there would be a traverse before the next fire bay.

Communication trenches between the two trenchlines tended to be zigzagged, and the second line would not necessarily need to have fire-steps, though it would probably still have fire bays and traverses. Second and third lines on German trenches would sometimes not be zigzagged but would curve. At this stage you can also think about the things you want in your trench system:

- Do you want to build a redoubt or strongpoint into the system?
- Are you going to build in machine-gun bunkers or *Stollen*?

- Where do you want the mortar pits?
- Do you want to have a sap pushed forward into no man's land?
- Where do you want the command posts and aid stations to be, and are you going to simply represent them or do you want to have them so that you can put figures into them?

If you are working on 60cm by 60cm boards, try not to have too many features on the same board or it will look overcrowded. Just one or two should be fine, and you can then build up different features on each board. If you are using 120cm long boards you can afford to have a few more on each board. It's the same with communication trenches. On 60cm boards you will really only need one or two or it will not look right.

Once you have decided where you want everything to be, draw it in but don't start to cut anything yet. Move the boards around again to see if you are happy with the location of everything and if you are not, now is the time to change it before any cutting has begun!

Scratch-Building

The scale you are working in, and how your figures are based, will also have a bearing on how wide you want your trenches to be. If you have multiple-based figures in 15mm then you may be happy to stick to having the bottom of the trench wide enough to put a base in comfortably and not worry about having a fire-step as it would mean that the base would sit in the trench at an angle. The same would be true for multi-based 20mm figures. However, if your figures are based singly, then you can consider having a fire-step on which the figures can stand. This should be just deep enough for one base to stand on, and you will then also want to have the right width in your trench to allow figures not on the fire-step to stand in it easily too. If you put two based figures one in front of the other, this should roughly give you the optimum width of the bottom of your trench.

Remember also that communication trenches tended to be much narrower than the front-line trenches with the fire-steps, so the bottom of these needs only to be about one base width (either multi-base or single figure). The same with second-line trenches; these may not have a fire-step so won't need to be quite as wide as your front-line trench.

Most trenches did not have straight sides, as this would have channelled any blast directly down the trench. There was no precise angle; it tended to vary on how and when the trench was built and who had been involved in digging it. Once you have worked out how wide you want the bottom of your trench system to be, add about 5mm each side on the width to allow for the angle. It might be best, if you can, to try this out on a spare piece of Styrofoam and see how it looks before you start cutting into your boards. It is a bit of trial and error to get the look right, so do make sure you are happy with your widths before you start cutting.

When you are drawing your trenches, it is sometimes best to draw two lines, the inside lines representing where you want the bottom thickness of your trench to be and a second set of lines each side of that showing where your top will be. You may find this helpful when you come to start cutting out your boards.

As you can carve Styrofoam, it is possible to hollow out each trench section down to the depth you would like them to be and to leave a small amount of Styrofoam at the bottom of the board. However, this will make the board quite fragile and easily broken. The best way is to mount each Styrofoam board on a square piece of MDF, which will give it strength and rigidity and support the Styrofoam.

You will need to make sure that the edge of each MDF board is straight or the boards will not butt up against each other neatly when you come to put them on the table. One way to do this is to get the boards cut for you at a DIY store, or a timber yard if there is one near you.

Once you have done this, you can start to cut out each piece of Styrofoam. Cutting the pieces is something of an art form, and is best done with your knife blade extended so that you can carve it. You could also cut them with a hot wire-cutter, but when you are doing this *PLEASE BE CAREFUL*!

Take your time and cut slowly and deliberately. Try to cut the pieces out along the inner (bottom) lines, so that you can then carve out the trench sides. You can either cut out the pieces and mount them on the board using solvent-free adhesive at this stage, or you can also carve out the trench sides before mounting them. I would probably choose the latter as it means you can handle the pieces more easily rather than trying to do it while they are glued down to the board.

Cut out one section at a time and then glue it in place. Once you have cut out the second section, make sure it fits into the place you want it before gluing it down, although solvent-free adhesive does give you a little bit of wriggle room before it dries permanently. Do one board at a time, moving on to the next board once you have one completed. Once you have a couple of boards done, try to match them up to make sure that the trenches meet at the edges where you want them to. If there are some small discrepancies you can correct these now with some careful carving, but please be careful not to carve too much off the Styrofoam as you can find that it will fit in that one place but may not then be interchangeable with anywhere else, which defeats the whole purpose. Keep all the off-cuts of Styrofoam as you may find that these come in handy later.

Once you have done all the trench boards and are happy that they all line up, you will want to check that they all sit well together and the edges are a good fit. If they are not, the problem might well be that the edges of one of the baseboards are not square. If this is the case you might have to gently shave some off until the boards fit together without an obvious gap between them.

If you are going to be transporting your boards regularly, perhaps either to a wargames club or to wargames shows, it is a good idea at this point to reinforce each corner. Styrofoam can be quite fragile if knocked or bumped in transit, and once a corner has broken off there's not very much you can do about it. I have found that a simple strip of tape on each corner helps to protect it when travelling about.

At this point, if you are making them, you could now do your no man's land boards. It might be tempting to cover all of the boards in shell-holes as this is the popular concept of no man's land. In fact a lot of the shell-holes would be close to one or other of the trenchlines, as artillery didn't often waste ammunition mounting barrages on an empty no man's land. Of course, if an offensive had taken place, then there certainly would be some shell-holes, but on our boards I want to limit this a bit so that you can easily put figures on the board. Additionally if the board is covered in shell-holes there will be no flat spaces to put extra scenery onto the board and it will make adding barbed wire later much more difficult.

The easiest way to make shell-holes is to first draw them in place, just as I did for the trenchlines. You can do this using something round that looks about the right size, such as the top of a crisps tube, but you will also want to vary the sizes a bit to represent different calibres of shell. You could even have two or more shell-holes together with their edges touching. Once you have drawn in a few, put the boards on the table together to see how they look. It might be best at this stage to mount the Styrofoam boards onto MDF. Even though you are not carving out trenches, they will need to be on the same level as your other boards and the shell-holes will also weaken the structure of the board without some support.

When you are happy that your shell-holes look how you want them to, you can start to carve them out. The easiest way to do this is with a chisel, but again *please be careful*. You can carve out chunks of Styrofoam from the board, but make sure that you put these to one side as you will need them. There isn't as much need to be regular and even; after all, these are holes that have been blasted into the ground. When you are digging out the Styrofoam, try not to make the shell-holes the same depth; you don't want them all looking identical.

Once you have carved out your shell-holes, break up the Styrofoam pieces that you have carved out into little chunks and glue them around the edge of the shell-holes using solvent-free adhesive.

Scratch-Building

Also don't forget that you don't just want shell-holes out in no man's land. The target for all the shelling was usually the enemy trench system, so you will need to dig out some shell-holes around the trenches on your trench boards. Exactly the same principles apply, though you should try not to make any shell-holes too close to your trenches as it might weaken the trench sides (which was of course the intention in reality!).

One of the other features that you might want to add to your no man's land board is a sap running forward from the trenchline. You will have thought about this when you were designing your trenches, and now is the time to build it in to your no man's land board. The only problem really is that once you have carved the sap in place, that no man's land board is in effect forever paired with the trench board that has the sap leading out from the trenches on it. If you are happy with this, then all is fine; if not, you might want to think about just having a small sap leading out from the trenchline and not all the way out into no man's land.

If you are carving a sap, you don't necessarily need to make it as regular as a trenchline. Saps were often dug quite quickly and usually at night. They generally had a position at the end that could be manned by one or two figures, or perhaps a machine-gun team. You could even build a sniper's position into one, although it would be a bit of an obvious place for a sniper. You won't need to build a fire-step though.

Once you have done this, it's time to go back to our trench boards. It's probably best to start doing the floors of the trenches. This is really just a matter of cutting out lots of pieces of coffee-stirrer and gluing them in place with solvent-free adhesive. Most floors went from side to side across the trench, but some German trenches had them running along the trench bottom. If they are going from side to side try not to put them too close

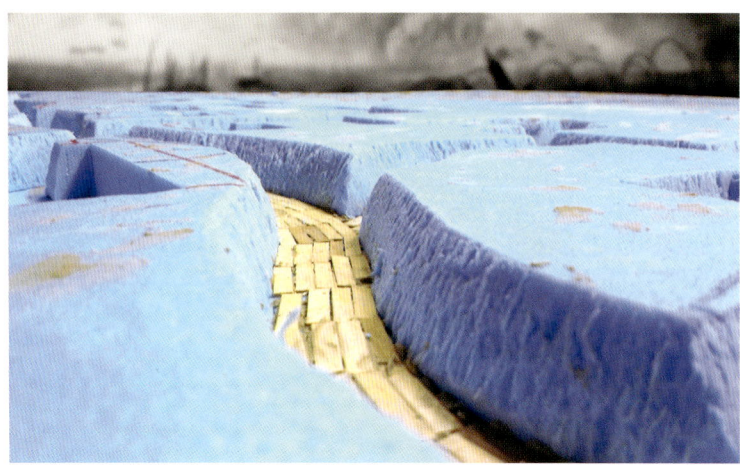

together; allow about a width of a stirrer between each one. If you have any actual dugouts, it's probably best not to put a floor in them as the thickness of the coffee-stirrers can make the difference between being able to put figures inside and still put the roof on and not being able to.

With the sides all carved out and the floors in place, you can now start to add some detailing to the sides of the trench. The first step is to decide where you want any dugout entrances (see what I said in the chapter on building 28mm terrain about this) and draw them in place. The next step is to start building the retaining walls on the trench sides.

For this you will need a *lot* of matchsticks. You can buy these in bulk quite easily in hobby shops or online. Please don't go out and buy boxes of matches and strike each one before cutting the heads off; there really is no need! Start by gluing a matchstick roughly every 3cm or so along the trench walls, making sure that you have one at each corner where the trench turns or joins. It doesn't matter if they are not exactly spaced all the way along; in fact that will make it look more realistic. Depending on the depth of your trench, you may have to cut them down so that the top of the matchstick is roughly level with the top of the trench. Make sure that you glue them with solvent-free adhesive.

Once these have set, you can start to add in the retaining walls. In the same way that I described in the section on detailing Amera trenches above, you can cut lengths of coffee-stirrer or string to fit between the matchsticks on the trench walls. Don't worry if the lengths are not all even as you are not after a totally regular look. Many sections of trench were blown in and then quickly rebuilt, so it would not all have been perfect. Trenches were frequently built in a hurry, so an uneven look is more realistic.

You don't necessarily need to cut and glue the strips one at a time either as that would take forever. The quickest and easiest way is to cut a batch and then glue them in place; if some of the pieces are too long then use them somewhere else or cut them down to fit. You can vary this by adding in some pieces of wrinkly tin in a few places, and by leaving some of the corners blank so that you can fill them with sandbags later if you want to.

Also don't forget to do the edges of the fire-steps. You will need to cut small pieces of matchstick for these and fill the gaps in between with wrinkly tin, which was the most common thing for supporting fire-step edges.

Scratch-Building

One of the features that you might want to add to your trench system is a blown-in section of trench. To do this you can carve out a chunk from the side of the trench (it's easiest if you do this from the back of the trench rather than the front) just as you have done for the shell-holes. When you come to do the retaining walls you can add to the effect by breaking the matchsticks and any coffee-stirrers you have used rather than cutting them to give a shattered look to it.

If you have any dugout entrances, you should edge these with matchsticks so they look a bit like the entrance to a Wild West coalmine! Place one matchstick either side of the entrance and one across the top. If they are 'false' entrances (in other words you don't have a dugout cut out on the other side), you can enhance the look by adding some painted tissue paper to represent the gas curtain every dugout had across its entrance.

A word of warning: this process is very laborious and tedious. It will feel like you are doing it forever, but once it is done the effect can be very realistic and you'll be glad you did it!

One of the things that you can add to German trenches is the A-frame support that sometimes went across the top of the trenches. Not all German trenches had this, but it does look really good if you can do a section of trench in it. Depending on the width of your trench, you can try to do this with matchsticks, but if they are not long enough then you may have to resort to balsa strip. Just cut a length so that it will go across the trench and glue it in place, perhaps for every other trench support piece.

Scratch-Building 87

If you have decided on any storage or shelter areas, now is the time to glue some wrinkly tin in place to act as a roof. You could also do this if you have built a latrine into the second-line trenches.

If you have a dugout into which you want to put figures, you will need to make the matchsticks that support the dugout sides slightly shorter so that they can support the roof when you put it in place. You'll need to work out what the thickness of the roof is likely to be – a piece of 3mm foamboard is good for a roof – so that it will fit smoothly in place when you fit it on.

You can use these as company command dugouts or perhaps forward aid stations.

Scratch-Building

Another feature you may want to add at this point is to include some machine-gun bunkers or some troop shelters, known by the Germans as *Stollen*. You could use resin bunkers from one of the many manufacturers or build them yourself from foamboard or Styrofoam. As these were mainly square or rectangular structures they are straightforward to build: just four sides and a roof.

By now you should be at the stage where all your walls and floors are completed. The next step is to cover all the tops of the boards in textured paint. This will take a while, and you will need to allow time for each board to dry thoroughly. Ideally this will be somewhere warm, possibly even laying them in the garden if it is a nice sunny day. Making boards like

Scratch-Building

this on cold or damp winter days, perhaps in a shed or garage, isn't really recommended as they will take a very long time to dry out completely.

You only need to cover the top of the boards; you don't need to put any of the textured paint into the inside of the trenches where you have done the walls and the floors. You will need to put some in the saps where there aren't any supporting walls or planked floors, and if you have done an area of collapsed trench.

When you are doing the areas around the shell-holes, be careful to cover all the little pieces of Styrofoam that you broke up and glued around the edges. You might need a bit more textured paint here to cover the pieces fully, but be careful not to put too much on or it will take a long time to dry out.

Once all the boards have dried, go back to them and look for areas where the textured paint hasn't covered as well as you thought it might. They will be pretty easy to spot! Just put another coat of the textured paint over where it's needed and let that dry fully too.

You should now have a set of boards that look like trenches in the winter! So our next step will be to paint each board fully, including inside all the trenches, with the dark brown emulsion. As well as the tops of the boards you will need to make sure that you also paint each of the sides, so that when they are all butted together you can't see any trace of the original colour of the Styrofoam.

This also needs to be left to dry thoroughly. Once it has, you will need to go round each board, probably several times, looking at it from all sorts of angles to find all the little unpainted white holes that you were sure you covered last time. There will inevitably be some, possibly lots, and if you don't paint them they will shout out to you every time you look at the board. Check especially around the areas where you had to put the textured paint on more thickly, such as the shell-holes and perhaps the saps too. It's probably best to use a smaller brush so you can get some paint right into the holes, but don't use your best brush as this will almost certainly ruin it.

Once you have painted all the boards using your dark brown emulsion and it has dried, you can now dry-brush them with the muddy colour. You can use quite a big brush for this where you are dry-brushing the tops of the boards, and a smaller brush to do the trench walls and floors. Where you have used coffee-stirrers or string, try to brush against the grain so the brush picks up on it. This should dry fairly quickly as you are only applying a light coat. When you are dry-brushing the shell-holes try not to put too much in the bottom of them, just a really light coat so it will add to the impression of depth.

The final jobs on painting the boards themselves are a light dry-brush over the tops of the boards and any earth areas using the caramel or wooden colour. Again this should be quite light as all you are trying to do is to pick up on the raised texture to highlight it. Lastly, take the silver-grey colour and dry-brush the matchsticks, planking and string areas to give it the impression of old wood. If you are not happy with the result just dry-brush back over with the muddy brown colour when it has dried and try another colour.

The next things to add are the sandbags. This is another job that part of the way through you will wish you had never started, but believe me it will look great when it's all done! It is exactly the same process that I described in the chapter on improving the vac-formed plastic Amera trenches, although of course you will need to adjust the size of the sandbags that you are rolling out to the scale of the model you are making. Again, don't worry if every sandbag isn't exactly the same size; once they are glued in place this won't be too noticeable. You will want to put a couple of rows along the top of the parapet and perhaps also the parados, and around any areas where you have put things like wrinkly tin for shelters. Once you have glued all the sandbags in place and they have all dried, paint them as usual with dark brown emulsion, then a heavy-ish dry-brush of Vallejo Desert Sand and lastly a dry-brush of Vallejo Iraqi Sand as we did before.

Once the sandbags are done and in place, we can add the barbed wire. Most trenchlines were protected by some form of barbed-wire entanglement, though this varied throughout

Scratch-Building

the war and even by nation. Initially barbed wire was erected on wooden posts but these had to be hammered into the ground, and the sound of troops doing that at night inevitably attracted fire or an enemy patrol. The British came up with a metal post that could be screwed into the ground, many of which can still be seen rusting away in museums all over the Western Front.

In modelling terms, these screw-in posts are not easy to represent. EWM do one in 20mm I think, but although its fine for trench stores you couldn't really use it to put into the ground and attach barbed wire to as it's not strong enough. One solution would be to use dressmaker's pins. Cut their heads off with a strong pair of pliers (but watch your eyes as if you are not careful the heads will ping everywhere, and you must find them as they will be a hazard) and then carefully file down the ends so they are no longer sharp. If you don't do this, make sure your tetanus injections are up to date as you will be stabbing yourself on the barbed wire every five minutes! As an alternative, if you are modelling in 20mm or 28mm, you could use cocktail sticks if you don't mind the wooden post look. Less lethal certainly, though you will have to paint them. Plastic rod would do a similar job.

Once you have your collection of pins or sticks, push them down through the textured paint and into the board. If you are using pins, make sure you use pliers to do this so the pins don't stick in your fingers rather than the board. Most barbed wire was at a height of around 4ft, so you should leave enough of the pin above the surface so that it looks about the right height with the scale of figures you are using, and don't forget that they will be based so you'll need to allow for that too. Try not to put them in straight lines but in a random order, about 10cm or so apart. Most barbed-wire entanglements were a few yards in front of the trench they were defending, so it's up to you whether you put them onto the edge of the trench board or the edge of the no man's land boards; both look good. Each barbed-wire system had places where troops could exit out into no man's land to patrol or raid the enemy trenches, so you'll need to allow for some clear lanes to do that without making them too obvious. These gaps were very often covered by machine guns too. If you are using pins, once you have them in place run a rust-coloured wash over them to disguise the metal of the pins.

You can now wind the barbed wire around the pins. A lot of manufacturers make scaled barbed wire, and it usually comes in lengths of a metre or so. Depending on how many boards you have you will need quite a lot of it. I would guess at roughly a metre for each 60cm board you are doing, but it will vary. To achieve a look of an entanglement, start at the middle of a section and twist the wire around one of the posts, and then secure that with a

blob of superglue. Let that dry, and then move on to the next post, twisting the wire around the post and securing with superglue as you go. Try to make it tight, but not too tight as the wire did sag over time in any event, and don't go in obvious straight lines; try to go in different directions but still making sure every post has wire attached to it.

When you were putting the pins out you may have come across a shell-hole. This would be accurate as one of the jobs of artillery was to try to reduce the enemy's barbed wire, one of the tasks that failed so spectacularly on 1 July 1916 (in part because a large proportion of the British shells were firing shrapnel, which doesn't cut wire very well). It looks quite good if you leave the pins at either end of the shell-hole, and when you are winding the wire round just leave a broken end to represent cut wire. Another thing you could do is to take the pins around the edge of the shell-hole so it looks like a wiring party has come out during the night and repaired the wire. Once you have your wire in place you could also run a wash of rust-coloured paint over it to take away the bare metal look.

A last finishing touch could be to add some flock to the middle of no man's land to represent grassy areas where there hasn't been much shelling, and to add some trench signs to each trench.

Chapter 6

PROJECTS

In this chapter we are going to look at making some terrain pieces that can either sit on your terrain or be built into your terrain as you make it.

Wargamers who were around in the '60s and '70s will fondly remember a brand of vac-formed terrain pieces by a manufacturer called Bellona. They were made to go with the figures that were then available from the likes of Airfix and Esci and so were usually scaled for what we would call 20mm figures today. They had some very useful pieces for the First World War including ruined houses and a terrain piece called the Menin Road which can make a great addition to the wargames table with a bit of work.

Sadly Bellona has been out of business for many years now and their terrain pieces are no longer available. However, from time to time some items do surface on eBay or on the bring-and-buy stalls at wargames shows and if you are using 20mm for your wargames armies, they are well worth collecting if you get the opportunity. I managed to obtain a few of the Menin Road terrain sets a few years ago. These are a reasonable size, and have the cellar of a ruined house alongside a section of road.

You can choose whether you want to integrate it into your scratch-built terrain or have it as a stand-alone piece that you can use as an objective or a defensive strongpoint. If you are building it into your terrain you could cut out a section of Styrofoam large enough to accommodate it and judge the depth so that it appears to be below ground level. You'll then need to build up the sides with textured paint to match it in. When built into a terrain board, it does look quite convincing.

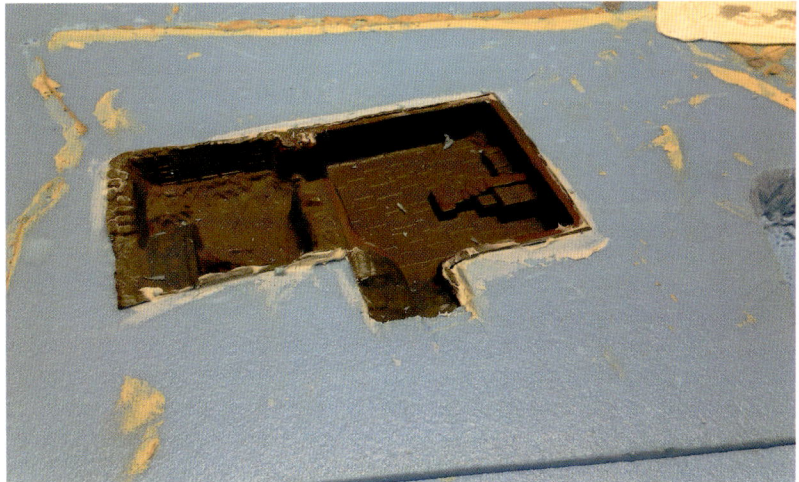

Alternatively you could just mount it onto one of your terrain boards and then blend in the joins using textured paint before you get to the painting stage.

If you are planning to have it as a stand-alone piece, you will probably need to base it with MDF as we did for the vac-formed pieces in the chapter above. This will give it some rigidity and prolong its life on the wargames table. You could then cover the outside in textured paint and paint it in the way we have described earlier or simply paint and drybrush it as it stands.

When making one of mine I chose to cut out part of the cellar section and build in a machine-gun bunker by Ironclad. The Germans in particular made great use of cellars for machine-gun positions and the bunker fits in there quite nicely. I finished it off with some sandbag protection made from air-dried clay as before and some coffee-stirrer planking to cover the gap.

WARGAMES TERRAIN & BUILDINGS

Projects

Another smaller Bellona piece was a ruined house. This could again be used either as a built-in or stand-alone terrain piece, and I decided to have mine as a stand-alone one. Again I mounted a machine-gun bunker in the house to form part of a defensive position.

Both strongpoints in use in a game.

Bellona also made a range of dug-in positions and even a trench section. These are a bit harder to find now, though if you do come across them they could easily be used on the wargames table with the techniques we have described.

Another of the iconic images of the First World War is the picture of troops in Passchendaele moving through a shell-shattered wood on duckboards. Areas of woodland featured on many of the battlefields of the Great War and their names still ring through history today: Polygon Wood in Flanders, High Wood on the Somme, Belleau Wood where the Americans had their debut in battle and so on. So for many modellers and wargamers including woods on the model battlefield is a must.

Luckily these are pretty easy to build. You can buy model 'dead trees' from companies such as Woodland Scenics and these do look pretty good, but a much easier and far cheaper way is to simply go out to your nearest stretch of woodland and collect some useful-looking twigs. These can be split and snapped at the top to represent damage, which after all is what happened to the real trees.

If you are building sections of woodland to place on your wargames table, the best way to do this is cut a shape out of MDF and then drill some small holes where you want the twigs to go. You may find that it is easier to cut the base of each twig so that it is reasonably level and then simply glue each twig where you want it to go. To add to the realism you could also glue some resin shell-holes, which several wargames scenery manufacturers make, around the board. Then, as we have done before, simply cover in textured paint and paint as usual.

If you are making bespoke terrain from Styrofoam boards, including a shell-shattered wood on your board is even easier since the twigs will easily press down into the terrain. If it is a board that you will be regularly transporting, you might want to make it so that they are removable to avoid damage (as they will catch on everything and anything). If, however, your terrain isn't going to be moved around very much, it is just as easy to fix them in place with some solvent-free adhesive. Trenches winding through the remains of a shattered wood can look very effective.

Another possible feature that could be added to a wargames table are some gun emplacements. In reality, of course, artillery would be a long way behind the front lines, often a mile or more, depending on the size of the gun and its range. However, some wargamers like to have the big guns on their table, and perhaps in 1918 games artillery emplacements could feature in a game with attacks by German stormtroopers or, conversely, by British, French or American infantry if it is after August 1918.

Projects

As with other scenery items, you can do this either by building the emplacements as stand-alone items or by digging them into the wargames board. There are several companies making gun emplacements including vac-formed ones, and these can be detailed using textured paint and sandbags in the way we have done for other items. You could also add some props from cocktail sticks and add some camouflage netting over the top of the emplacement.

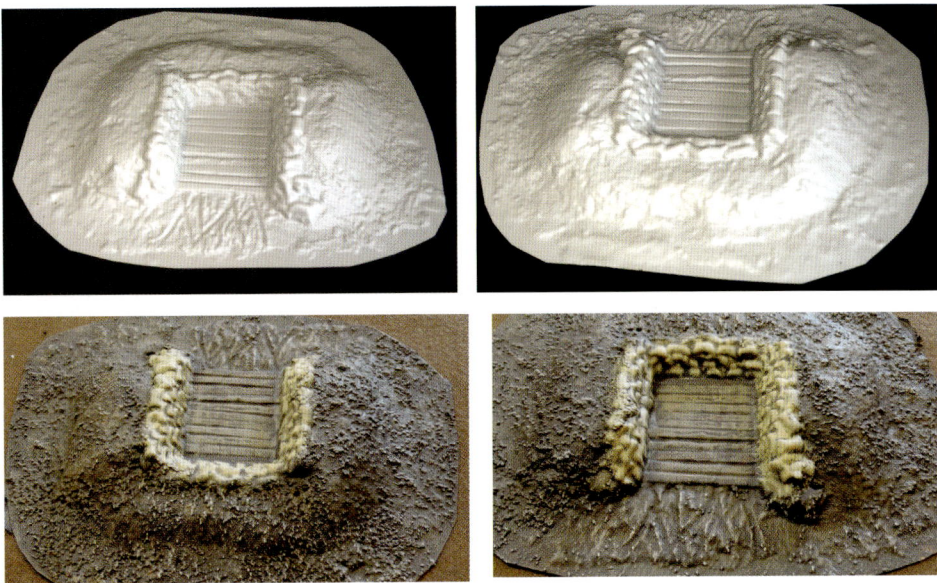

In 15mm or 28mm, you could use one of the artillery positions made by companies such as Ironclad. They can be incorporated into a small defensive position using some of the trench pieces.

If you want to make dug-in positions on your wargames board, this can be easily done by carving out the shape for your gun on the Styrofoam and then texturing as usual, adding some coffee-stirrers to represent the base for the artillery piece and some sandbags.

Chapter 7

RESEARCHING FIRST WORLD WAR TRENCHES TODAY

Whatever you are planning to build, one of the first things to do is to research what the trenches actually looked like. There are many ways that you can do this; for example, by looking at books, at photographs and other resources online and if you can by visiting the battlefields themselves.

A great many books have been written about the trenches and what life was like in them. For wargamers, a good place to turn for inspiration is the excellent series of books by Osprey Publishing. There are several books that will be very helpful, such as *Trench* by Stephen Bull (which seems to be a couple of other Osprey books joined together), *Fortifications of the Western Front 1914–18*, *The Hindenburg Line* and the two-volume *WW1 Trench Warfare*. These are a pretty good start for anyone wanting to research how the trenches looked and evolved over the course of the war.

There are, of course, almost endless publications about the Great War, many of which will have photographs of trenches and trench systems in them. The classic picture, often used in books on the Great War, is of the British infantryman leaning with his rifle and bayonet on the side wall of a trench on sentry duty. However, if you look more closely at the picture, you will see that it is in fact a German trench that has been 'turned round' so that the parapet has now become the parados. This does, however, give quite a good picture of a trench in the middle stages of the war.

If you can get hold of a book that has aerial photographs of the trench systems, that will give you a fair idea of the general layout and shape in the landscape that the trenches took. Books of trench maps are also available, and these can give a very clear idea of the way the trenches spread over time. *The Western Front from the Air* by Nicholas Watkis is a good example of this, with some useful photos of the trenches as they spread across the landscape.

We are so incredibly fortunate today that the number of photographs from the Great War available online is prodigious. A simple search with your search engine of choice should quickly bring up as many photographs as you are likely to need. A great example of this is the photographic library of the Imperial War Museum, which is an excellent free online resource providing loads of photographs of the trenches on almost every front in the war. These should give you plenty of inspiration for making your trench models look as realistic as possible.

Visiting the Battlefields

While books and photographs can give you a good idea of what the trenches looked like, they are no substitute for walking the ground if you possibly can. The battles of the Western

Front in Belgium and Northern France are within easy reach if you happen to live in the UK or from the large cities such as Bruxelles or Paris if you are flying in. The area around Ypres is no more than an hour's drive from the ferry ports of Calais or Dunkirk (which is quite a bit nearer) or the Eurotunnel at Coquelles. It is about an hour and a half from Bruxelles airport too, so easily drivable.

A little further will take you to the battles around Arras, which again is easily accessible for a day trip from the UK. Calais would be your best port for this area as it's only about an hour's drive from there. Loos is another easy battlefield to reach; again this is only about an hour from the Channel coast.

To venture further afield, the Somme is just about reachable from the Channel ports to allow for a day trip, though the drive is about two hours each way. This does make for a fairly long day, and if you want to fully explore the battlefield it might be best to stay somewhere like Amiens. If you are flying in, Amiens is about an hour and a half's drive from Charles de Gaulle Airport near Paris.

Mons is also around two hours from the coast, so again could be done in a long-ish day trip.

Slightly nearer to Paris, and probably beyond a day trip from the UK, are the battlefields of the Marne. Any further than this, for example down to the 1916 mincing machine at Verdun or the Chemin des Dames, will require an overnight stay or a few days away.

There are plenty of options available for visiting the battlefields. In addition to doing it yourself, there are very many companies offering short guided trips to specific areas and these can often be one of the best ways to see the area. They offer a wide range of tours, from an overview of an area or battle, walking an area such as the Somme, or very detailed looks at particular areas often guided by well-known authors and experts on the battle. If you have not visited the battlefields before, this can be a very good way of learning about the battle and seeing the physical geography of the area. Perhaps the only down side of organized trips is that you will never get a very long time in any one area, though you will be able to visit some very good museums and locations, and this could be a good way of finding locations for later visits on your own.

While visiting the battlefields of the Great War is undoubtedly enjoyable, if you are specifically looking to recreate the war's trench systems in all their variety and complexity, your needs are a little more specialized. There are, however, several museums across the front that have either preserved or reconstructed trenches remaining to provide you with inspiration and to get those trench models completed!

Here are a few examples of those that are well worth visiting, most in the Ypres area and so within easy reach of the UK. There are others that I have not covered in detail, such as the Yorkshire Trench to the north of Ypres and the Essex Farm bunkers that formed the dressing station where John McCrae, who wrote the *In Flanders Fields* poem, was stationed in 1915. At Vimy Ridge near Arras there is a short section of the front line of both the British and German armies preserved in concrete, with a short stretch of no man's land in between. Then of course at Verdun there are the massive preserved forts of Douaumont and Vaux.

Trench of Death

Working from north to south, the first example to mention is the Trench of Death at Diksmuide, which is between Nieuwpoort on the Belgian coast and Ypres. If you drive on the coastal autoroute from the Channel towards Belgium you will come across several signs to Diksmuide not long after you have crossed the border with France and the trenches are located just outside the town. They are signposted from the town centre, and are called *Dodengang* in Flemish and *Boyau de la Mort* in French.

These are the remains of the Belgian trenches on the banks of the Yser Canal, where a bitter war was fought for almost four years across the flooded fields in the only remaining part of Belgium that was not occupied by the Germans. Given the nature of the fighting here, their name is probably very appropriate.

The trenches have been preserved in concrete and there is also a small museum on the site, as well as a German bunker. The trench system stretches for about 300 yards or so, and has firing positions with traverses facing across the river and a supervision trench running immediately behind the firing line. In common with many trench systems in this part of Flanders, it is built above ground level rather than being dug in.

This is a good example of a trench system set up in late 1914 when the Battle of the Yser was taking place. The trenches then evolved over time, as the war did not move at all on this front until September 1918 when the Belgian army advanced along with the other Allied armies.

Trench of Death, Diksmuide, Belgium.

Dugout entrance.

Machine-gun position or lookout.

Position overlooking the Yser Canal.

Memorial Museum, Passchendaele
Moving down towards Ypres, the next and possibly best museum for getting an idea of what the trenches looked like is the Memorial Museum Passchendaele at Zonnebeke Château.

This excellent museum located on the Passchendaele Ridge not only has an extensive and fascinating collection of exhibits and artefacts in the museum including a lot of uniform dummies that are great for painting reference, but more importantly for us it has a large reconstructed trench system in the grounds depicting both British and German trench systems at different stages of the war. There is also a large underground bunker with a communications centre, an aid post and a headquarters area. All of these are populated by uniformed dummies to give a realistic impression of what life was like below ground. The main theme of the museum is naturally the Third Battle of Ypres in 1917.

You do need to remember that these are reconstructed trenches, however. There were undoubtedly trenches on this site, but the museum has taken the (very helpful) decision to try to recreate what the trenches would have looked like, broadly speaking, when they were in use. This does give a really useful insight into how they were constructed and what the trenches we are trying to reproduce in miniature should look like.

Part of a traverse leading to a firing position. You can see the way in which the trenches were wider at the top than at the bottom, to help channel blast.

This trench doesn't have a fire-step or traverses, so is most likely an example of a communication trench.

A British firing step, with the front reinforced with chicken wire.

Another view of the firing step. You can clearly see the duckboards running along the bottom of the trench here, and the wrinkly tin supporting the trench and fire-step sides at the bottom.

This picture gives you a good idea of how a firing bay with traverses would have looked. The sides are protected with sandbags but could just as easily have been boarded, with more use of wrinkly tin if it was available. In the early days of trenches in 1914 trench sides might have been reinforced with anything the troops could find, such as doors from inside nearby houses.

In contrast to the British firing bay this is an example of a German one, although a French one might also have looked similar. You can see the use of wattling to reinforce the trench sides. The sandbags here are slightly higher, with firing loopholes.

Researching First World War Trenches Today 111

Another view of the trench, with the firing step going round into the traverse. In this trench the traverse seems to be at a 45-degree angle rather than 90 degrees.

An overview of the position. The wall between the two trenches looks very thin and unlikely to survive a blast!

An example of the inverted A-frames used to support the German trenches. These could sometimes be roofed over to provide more cover, though not much protection.

A closer view of the inverted A-frames.

Researching First World War Trenches Today

A sniper position with a metal shield.

A trench shelter, known to the British as an elephant shelter and the Germans as a Heinrich shelter. This is a German example.

Inside the German shelter.

The British version of the same thing.

Above ground. Very little can be seen.

Hooge Crater

To the east of Ypres, along the Menin Road near the Bellewaerde Aquapark, is the Hooge Crater Museum, cemetery and the Hooge Crater itself. This was the location of the first use of flame-throwers by the German army against the British in 1915, and the crater itself is the result of a mine explosion by the British, also in 1915. The Hooge Crater Museum is an excellent private collection of First World War artefacts and well worth a visit, though there is not much in the way of preserved trenches on the site. The Aquapark stands on the site of Hooge Château, which was destroyed during the war.

The crater itself is a little further back on the road towards Ypres in the grounds of a hotel, and you can visit it by entering a small gate to one side (when we went there was a voluntary contribution payable on entry). Here you can see the remains of some original trenches and a German bunker. The trenches were first built by the Germans in early 1915 and in July 1915 they were captured by the British, including blowing a mine underneath the German positions which is what created the crater that remains today, filled with water. They lost them a few days later in the first flame-thrower attack but then regained them again.

The trenches themselves are relatively shallow and the sides are supported by wrinkly tin. They do, however, follow fairly closely the location of the original trenches and so give a good idea of the trench layout on this site.

Trenches covered with elephant shelters and sides supported by wrinkly tin.

Good simulation of a walkway over a trench.

Researching First World War Trenches Today

The trenches weave their way through what is now a small wood.

Sides supported by wrinkly tin.

WWI TRENCH SYSTEMS

The trenches are fairly shallow but give an idea of how it would have been at the time.

Wrinkly tin used as a shelter over the trenches.

Researching First World War Trenches Today

The path down into the trenches.

German bunker entrance.

Remains of German bunker and path leading round to the trenches.

Hill 62 and Sanctuary Wood

A little further round is the Sanctuary Wood Museum located on Hill 62. This is a very interesting museum, with displays of all sorts of First World War-related equipment and weapons, and at the back of the museum is a section of preserved trenches.

It is believed that the wood that was originally on this site, to the south of the Menin Road, was used in 1914 by the British army as an area to treat their wounded and so became known as Sanctuary Wood on trench maps. The trenches in this area were established at the end of 1914, and remained in use here until the Third Battle of Ypres in 1917. After the war the farmer who owned the land left this section of the British trenches largely as he found them and they became a preserved trench system. The family still own the land and the museum and the trenches mainly follow the line of the original ones, although they have been restored over the years.

A good impression of the layout of a front-line trench with fire bays and traverses. The trench sides have been supported with wrinkly tin in the same way as at Hooge Crater.

Researching First World War Trenches Today

Trenches zigzag through the wood, most likely communication trenches.

A good view of the lines of the trench leading through the wood.

The remains of a trench shelter, with genuine Flanders mud!

Another view of the fire bays and traverses.

A dugout/bunker with a stone-built entrance.

Trenches leading round to the bunker entrance.

Remains of shell-holes in the wood.

Remains of a sniper shield.

Bayernwald (Bayern Wood)

To the south of Ypres there is Mont Kemmel or the *Kemmelberg*, which rises up to a height of 156 metres. Not far from Kemmel village are the German trenches known as *Bayernwald*, though the area was known as Croonaert Wood to the British. The area was discovered in 1971 but fell into disrepair until it was restored in 2004. The trenches are not the originals, but follow the lines on German trench maps.

This is another trench system with the sandbags preserved in concrete and includes extensive trenches and several bunkers, though these appear to be *Stollen* shelters rather than machine-gun bunkers. You do need a ticket to enter through the turnstile entrance, which can be purchased from the Belgian Tourist Office in Kemmel village. One of the interesting facts about this trench system is that it is believed that Adolf Hitler was stationed here for a while in 1915 and served as a runner.

The trench system gives you a really clear idea of what a German trench system would have looked like, and is well worth exploring if you are in the area and have the time.

German trenches with concrete sandbags and wattle sides. The trenches curve in a wavy line rather than zigzagged fire bays and traverses.

A closer view of the trench sides.

An entrance to a tunnel leading down to an underground bunker.

A useful view of a trench junction.

Researching First World War Trenches Today

The entrance to a concrete bunker, most likely used as a shelter.

Another bunker entrance.

The other side of one of the bunkers and the trench leading away.

Another view of the trench system.

Auchonvillers

One of the more interesting sections of preserved trenches on the Somme is at Auchonvillers. These are located to the rear of a very popular café and hotel in an area better known to the British at the time as 'Ocean Villas'. The café is located near the Newfoundland Park.

The trenches were originally dug by the French in 1914 as a communication trench leading to and from the front line. It was taken over by the British in 1915 and was used as a communication trench in the days leading up to the Battle of the Somme, and on the eve of the 1 July battle itself in the attack on Beaumont Hamel.

The trench sides are again preserved by using wrinkly tin.

A short stretch of trench leads alongside the building and is covered by a shelter.

Visitors are often observed by bemused sheep!

The trench protected by shelters.

Researching First World War Trenches Today

A narrow section of trench.

Looking back along the section of trench.

Trench entrance.